Ronald Breslow
Artificial Enzymes

Further Titles of Interest

A. F. Collings, C. Critchley (eds.)
Artificial Photosynthesis

2005. ISBN 3-527-31090-8

E. Keinan (ed.)
Catalytic Antibodies

2004. ISBN 3-527-30688-9

A. S. Bommarius, B. R. Riebel
Biocatalysis

2004. ISBN 3-527-30344-8

K. Drauz, H. Waldmann (eds.)
Enzyme Catalysis in Organic Synthesis

2nd Edition, 2002. ISBN 3-527-29949-1

S. Brakmann, K. Johnsson (eds.)
**Directed Molecular Evolution
of Proteins**

2002. ISBN 3-527-30423-1

Edited by Ronald Breslow

Artificial Enzymes

WILEY-VCH

WILEY-VCH Verlag GmbH & Co. KGaA

Editors

Prof. Dr. Ronald Breslow
Columbia University
Department of Chemistry
556 Chandler Laboratory
New York, NY 10027-6948
USA

Library of Congress Card No.: Applied for.

British Library Cataloguing-in-Publication Data:
A catalogue record for this book is available from the British Library.

**Bibliographic information published by
Die Deutsche Bibliothek**
Die Deutsche Bibliothek lists this publication in the Deutsche Nationalbibliografie; detailed bibliographic data is available in the Internet at http://dnb.ddb.de

© 2005 WILEY-VCH Verlag
GmbH & Co. KGaA, Weinheim

Printed in the Federal Republic of Germany

Printed on acid-free paper

Cover Design Gunter Schulz, Fußgönheim
Typesetting Mitterweger & Partner, Kommunikationsgesellschaft mbH, Plankstadt
Printing Strauss GmbH, Mörlenbach
Bookbinding Litges & Dopf Buchbinderei GmbH, Heppenheim

ISBN-13: 978-3-527-31165-1
ISBN-10: 3-527-31165-3

Table of Contents

Artificial Enzymes. Edited by R. Breslow
Copyright © 2005 WILEY-VCH Verlag GmbH & Co. KGaA, Weinheim
ISBN: 3-527-31165-3

Preface

Chemists have been extending Nature for hundreds of years. This has generally taken the form of creating new compounds that did not exist in Nature, including new medicines, new polymers, and simply new interesting compounds such as novel aromatics, for instance. At this point well over 90% of all known chemical substances are unnatural products. They were often developed by imitating the general style of natural substances – a good example is synthetic polymers, which differ from but are intellectually related to the polymers of biology. The motivation for extending Nature is often simple curiosity about what else is possible, but there is always the hope that a useful new property might emerge.

I have coined the term Biomimetic Chemistry to describe novel chemistry that is inspired by that done in living systems. In that sense modern aviation is biomimetic, inspired by the principles we see in birds and flying insects (wings) but not the exact details of how they are used in living creatures (we gave up on efforts to flap wings after some disastrous tries in early airplanes). Artificial Enzymes are part of the world of Biomimetic Chemistry, in which we are imitating the *catalytic processes* that occur in living systems, not just the substances of biology. We hope to achieve the rapid and selective reactions typical of enzymatic conversions by imitating the general principles – in particular the well-defined geometry in enzyme–substrate complexes – but not every detail of natural enzymes. This will greatly increase the power of synthetic chemistry.

There is another important function of biomimetic chemistry – increasing our understanding of natural chemistry by putting it in a larger context. Thus, the properties of artificial enzymes help us understand and appreciate the special properties of Nature's enzymes. In particular, the great challenge in achieving, with artificial enzymes, the huge rate accelerations that the best natural enzymes can achieve makes us think more deeply about what is special in natural enzymes. As one example, the work on polymeric artificial enzymes described in Chapters 2 and 3 indicates a special role that the macromolecular structures of enzymes can play in excluding water solvent from the reaction zone.

The seven chapters in this book describe various approaches to the synthesis and study of artificial enzymes. In Chapter 1, I describe work in my laboratory over the past almost 50 years creating enzyme models and enzyme mimics. A major theme is the use of hydrophobic binding of substrates into cyclodextrins carrying catalytic groups,

Artificial Enzymes. Edited by R. Breslow
Copyright © 2005 WILEY-VCH Verlag GmbH & Co. KGaA, Weinheim
ISBN: 3-527-31165-3

to imitate the reversible formation of enzyme–substrate complexes. In Chapter 2, Lei Liu and I describe a particular aspect of our and related work in more detail, namely the creation and study of artificial transaminases and other enzyme mimics based on co-factors in the pyridoxal/pyridoxamine family. In this work we have taken advantage of the pioneering studies, by Irving Klotz, of artificial enzymes based on ethylenimine polymers. Then, in Chapter 3, Klotz and Suh describe other work on these fascinating polymers, which show both the ability to hide a reaction zone away from water solvent and also the intrinsic effective acid–base catalysis by the amino groups of the polymers.

Since catalysis can be achieved by selective binding of the reaction's transition state (binding is simply an interaction between two components that lowers the free energy of the system), there is much interest in the use of antibodies as artificial enzymes – antibodies raised against structures resembling the transition states of reactions. When they bind the transition states, the reaction is accelerated. In Chapter 4, Hilvert describes the remarkable advances in this field, including his own ground-breaking creation of antibody-based artificial enzymes that can substitute in a living organism for the natural enzyme. Antibody-based artificial enzymes are, of course, proteins, and in Chapter 5 Duckworth and Distefano describe another approach to artificial enzymes – modifying natural proteins to impart novel enzymatic activity.

Many natural enzymes use metal ions for their principal catalytic functions, so artificial enzymes have been created using such metal ion catalysis. Chin and Kim describe this area, and their own exciting work, in Chapter 6. One of the areas in which natural enzymes play an important role is in molecular biology; restriction enzymes can hydrolyze DNA at specific target sites and facilitate genetic engineering. In Chapter 7, Yamamoto and Komiyama describe trail-blazing work on the construction of artificial restriction enzymes. They use metal ion catalysis, but with site selectivity to cleave both single stranded and double stranded DNA.

I hope that this book, a sampling of work on artificial enzymes, conveys the flavor and some of the achievements of the field. As in the past, chemists have not simply admired nature, they have set about to learn from it and extend it. The result is novel catalysts that can be tailored to achieve selective reactions previously available only in enzymatic processes. With such a promising start, I believe that the field of Artificial Enzymes, as part of Biomimetic Chemistry, will continue to grow into a major area of modern chemistry. The methods can be attractive in the synthesis and manufacture of medicinal compounds, and artificial enzymes themselves may become useful as therapeutics and in molecular devices. Thus I urge young students to join the effort to revolutionize chemical catalysis by applying the principles of enzymatic reactions to the invention of artificial enzymes, and their applications. There is still much to be done.

Columbia University
December 2004

Ronald Breslow

List of Authors

Ronald Breslow
Columbia University
Department of Chemistry
566 Chandler Laboratory
New York, NY 10027-6948
USA
rb33@columbia.edu

Jik Chin
University of Toronto
Lash Miller Chemical Laboratories
80 St. George St.
Toronto, Ontario M5S 3H6
Canada
jchin@chem.utoronto.ca

Mark D. Distefano
University of Minnesota
Department of Chemistry
207 Pleasant St SE, Smith Hall
Minneapolis, MN 55455-0431
USA
distefan@chem.umn.edu

Ben Duckworth
University of Minnesota
Department of Chemistry
207 Pleasant St SE, Smith Hall
Minneapolis, MN 55455-0431
USA

Donald Hilvert
Swiss Federal Institute of Technology
Laboratory of Organic Chemistry
ETH-Hönggerberg
CH-8093 Zürich
Switzerland
hilvert@org.chem.ethz.ch

Hae-Jo Kim
University of Toronto
Lash Miller Chemical Laboratories
80 St. George St.
Toronto, Ontario M5S 3H6
Canada

Irving M. Klotz
Northwestern University
Department of Chemistry
2145 Sheridan Road
Evanston, IL 60208-3113
USA
i-klotz@northwestern.edu.

Makoto Komiyama
The University of Tokyo
Research Center for Advanced Science
and Technology
4-6-1 Komaba, Meguro-ku
Tokyo, 153-8904
Japan
komiyama@mkomi.rcast.u-tokyo.ac.jp

Artificial Enzymes. Edited by R. Breslow
Copyright © 2005 WILEY-VCH Verlag GmbH & Co. KGaA, Weinheim
ISBN: 3-527-31165-3

Lei Liu
Columbia University
Department of Chemistry
New York, NY 10027-6948
USA

Junghun Suh
Seoul National University
Department of Chemistry
Seoul 151-747
Korea
jhsuh@snu.ac.kr

Yoji Yamamoto
The University of Tokyo
Research Center for Advanced Science
and Technology
4-6-1 Komaba, Meguro-ku
Tokyo, 153-8904
Japan

1
Artificial Enzymes

Ronald Breslow

Biomimetic Chemistry, including that involved in the synthesis and study of artificial enzymes, has grown to enormous proportions. Even the part of the field using cyclodextrins as binding groups in synthetic catalysts that mimic enzymes has been the subject of a large review article [1]. Thus in this chapter I will focus mainly, but not exclusively, on work from our own laboratory. Other chapters will help make up for this somewhat narrow focus. I have published several reviews of our work elsewhere [2–51].

1.1
Mimics of Enzymes that use Thiamine Pyrophosphate as a Coenzyme

I have been pursuing enzyme mimics, artificial enzymes that perform biomimetic chemistry, since starting my independent career in 1956. In the first work [52–59] my co-workers and I studied models for the function of thiamine pyrophosphate **1** as a coenzyme in enzymes such as carboxylase. We discovered the mechanism by which it acts, by forming an anion **2** that we also described as a stabilized carbene, one of its resonance forms. We examined the related anions from imidazolium cations and oxazolium cations, which produce anions **3** and **4** that can also be described as nucleophilic carbenes. We were able to explain the structure–activity relationships in this series, and the reasons why the thiazolium ring is best suited to act as a biological

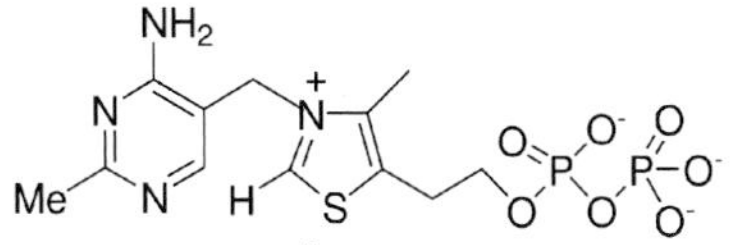

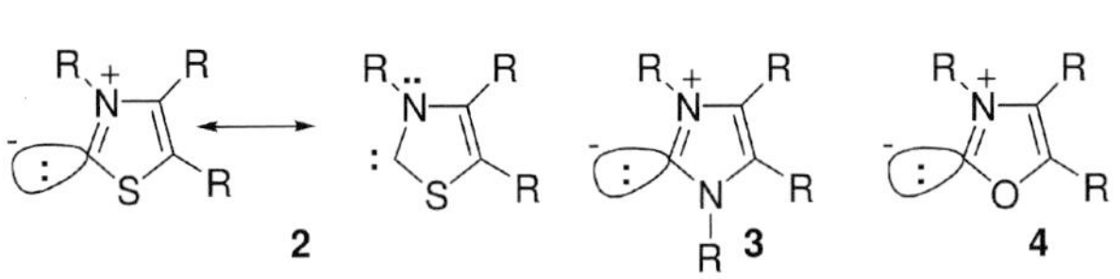

Artificial Enzymes. Edited by R. Breslow
Copyright © 2005 WILEY-VCH Verlag GmbH & Co. KGaA, Weinheim
ISBN: 3-527-31165-3

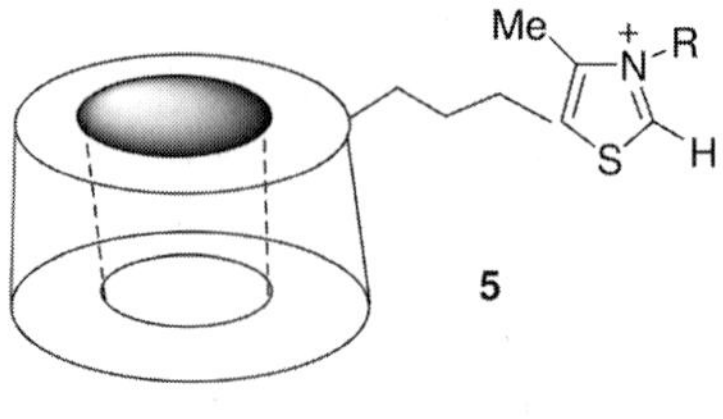

β-Cyclodextrin

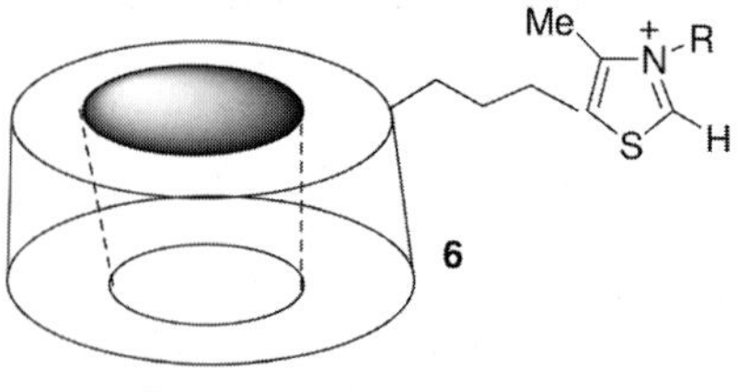

γ-Cyclodextrin

catalyst. Later, we confirmed [60] the thiamine mechanism proposed earlier, for which an alternative had subsequently been proposed [61, 62].

We synthesized artificial enzymes **5** and **6** that incorporated the thiazolium ring of thiamine into a cyclodextrin binding unit [63, 64]. The cyclodextrin imitated the hydrophobic binding pocket typical of many enzymes, but these mimics did not incorporate the catalytic groups that enzymes also use. Thus **5** and **6** showed the substrate selectivity that enzyme binding also achieves, and there was some rate acceleration from binding the substrates in proximity to the catalytic coenzyme group, as in enzymes. However, the rate accelerations were not nearly as large as those in artificial enzymes (vide infra) that incorporated more features of natural enzymes.

With β-cyclodextrin, consisting of seven glucose units in a ring, benzaldehyde bound into the cyclodextrin cavity of **5** and was converted into thiazolium adduct **7**, similar to a cyanohydrin [63]. This readily formed the benzylic anion **8** that underwent deuterium exchange and easy oxidation. However, this β-cyclodextrin ring was too small to bind both benzaldehydes – so this artificial enzyme did not catalyze the formation of benzoin **9** significantly better than did a simple thiazolium salt without the attached cyclodextrin binding group.

When the thiazolium unit was attached to the larger γ-cyclodextrin in **6**, with eight glucose units in the ring, benzoin condensation of two benzaldehydes was indeed well

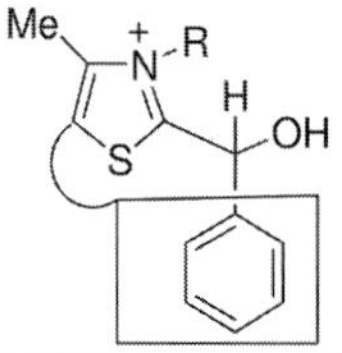

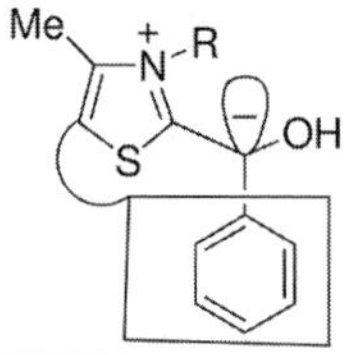

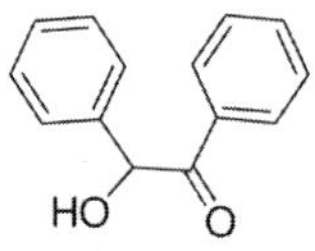

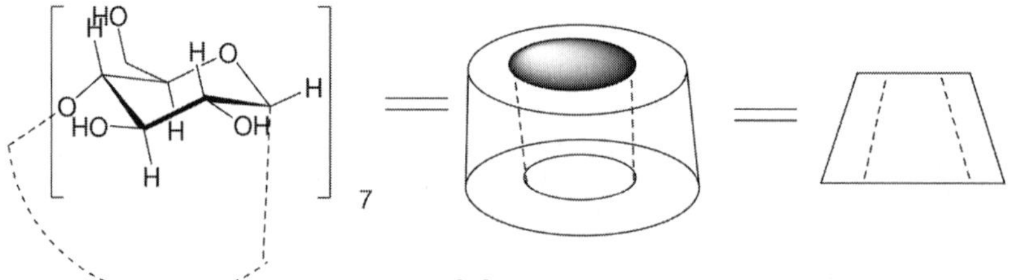

β-Cyclodextrin

catalyzed, with a rate 150-fold higher than that for a thiazolium salt lacking the cyclodextrin. Interestingly, in this benzoin condensation the rate-determining step – addition of anion **8** to the second benzaldehyde – allowed the benzaldehyde units to bind next to each other in the cavity, but in the product benzoin **9** the extended geometry does not permit this. Thus the benzoin product did not bind strongly to the artificial enzyme **6**, and did not inhibit the process. Our other studies on the benzoin condensation [65, 66] revealed geometries of the transition state and product that support this interpretation.

1.2
Mimics of Enzymes that use Pyridoxamine and Pyridoxal Phosphates as Coenzymes

We also attached pyridoxamine to a cyclodextrin and saw that the resulting enzyme mimics showed good substrate selectivity in the conversion of keto acids into amino acids [67–71]. With a pyridoxamine doubly-linked to the cyclodextrin there was a preference for the hydrophobic *t*-butylphenylpyruvic acid relative to pyruvic acid of at least 15 000-fold. We also made a related system, in which a synthetic macrocycle was attached to the coenzyme mimic [72], that also showed substrate selectivity. In other work we synthesized molecules in which base groups attached to the pyridoxamine could perform transaminations with good stereoselectivity [73–75]. We also made others in which the geometry of the attached base groups could promote different catalyzed processes for pyridoxal, selecting among the various enzymatic processes for which pyridoxal phosphate is a coenzyme [76–80].

These and subsequent artificial enzymes that perform transaminations are described in Chapter 2.

1.3
Artificial Hydrolytic Enzymes

1.3.1
Chymotrypsin Mimics

The field of artificial enzymes has been greatly concerned with mimicking *hydrolytic* enzymes. Since the enzyme chymotrypsin was one of the first to be extensively studied and understood, many laboratories have created artificial peptidases and esterases, including those that use the nucleophilic mechanism like that in chymotrypsin. (However, one chymotrypsin mimic from other laboratories did not have the reported mechanism [81].) The critical requirement is bifunctional catalysis, which in chymotrypsin involves imidazole acting first as a general base, then as a general acid, and the serine hydroxyl group serving as a nucleophile. I have pointed out the special kinetic situation this mechanism implies [82].

We have made several artificial enzymes that use cyclodextrin to bind a substrate and then react with it by acylating a cyclodextrin hydroxyl group. This builds on earlier work by Myron Bender, who first studied such acylations [83]. We added groups to the cyclodextrin that produced a flexible floor, capping the ring [84]. The result was to increase the relative rate of cyclodextrin acylation by *m-t*-butylphenyl acetate from 365 relative to its hydrolysis rate in the buffer to a $k_{complex}/k_{buffer}$ of 3300. We changed the substrate to achieve better geometry for the intracomplex acylation reaction, and with a *p*-nitrophenyl ester of ferroceneacrylic acid **10** we achieved a relative rate for intracomplex acylation of ordinary β-cyclodextrin vs. hydrolysis of over 50 000 and a V_{max} comparable to that for hydrolysis of *p*-nitrophenyl acetate by chymotrypsin [85].

Our best combination of the flexible capped cyclodextrin with the well-fitting substrate *p*-nitrophenyl ester **10** gave an acceleration – relative to hydrolysis in the same medium – of over one-million fold, exceeding that achieved by chymotrypsin with *p*-nitrophenyl acetate [86]. An even better fitting substrate (**11**) afforded an acceleration of ca. 80 000 000-fold, and saw a 62-fold increase in enantioselectivity as well [87, 88]. This is an enantiomeric excess of 98.4%.

Substrate binding into the cyclodextrin cavity, which ordinarily is studied in water solution, also occurs in highly polar organic solvents such as DMSO [89]. Furthermore, kinetic studies of our reactions at high pressure were consistent with the geometries proposed for these acylation processes [90]. Molecular modeling showed geometries of the bound substrates and the tetrahedral intermediates that helped explain some of the large rate effects [91].

In the acylation of a cyclodextrin hydroxyl group by a nitrophenyl ester, the preferred geometry requires that the oxyanion of the cyclodextrin attack perpendicular to the plane of the ester carbonyl, so as to form the tetrahedral intermediate. However, the product cyclodextrin ester has the cyclodextrin oxygen *in* the plane of the carbonyl group. Thus, a rapid reaction requires enough flexibility to be present to permit this geometric change to occur rapidly. With very rigid substrates the conversion of the tetrahedral intermediate into the product can be rate determining, and slow. In a study of this question we used substrate **12** in which the ester carbonyl can freely rotate, and saw that this made the formation of the tetrahedral intermediate rate-limiting, and rapid [92].

When functional groups are attached to the cyclodextrin ring, new artificial enzymes can result. We have already described this for the attachment of thiazolium rings and pyridoxamine/pyridoxal rings above, and will describe the attachment of metal catalytic groups in the next section. However, one study with attached phosphate groups addressed a general question: Is there a preference for putting such groups on the primary or secondary side of the cyclodextrin ring [93]? In the studies mentioned to this point, the catalytic groups were attached to the primary CH_2 groups of the cyclodextrins, but the acylation reaction occurred on the secondary CH–OH groups.

We prepared phosphate esters **13** and **14** of both the secondary hydroxyl and primary hydroxyl groups of β-cyclodextrin, and examined them as general acid and general base catalysts for the reactions of bound substrates [93]. The phosphate anion acted as a general base to catalyze enolization and the resulting hydrogen exchange in a bound tritiated phenacyl ketone **15** when the phosphate was either primary or secondary, showing that the substrate can bind equally well and undergo catalysis when pointed in either direction in the cyclodextrin cavity. However, the general acid-catalyzed hydrolysis of a bound acetal **16** had a preference for the phosphoric acid group on the secondary side. The position of attachment of the catalytic groups can be important.

1.3.2
Metalloenzyme Mimics

Chymotrypsin is only moderately effective as an enzyme, and much higher rates are seen with metalloenzymes. Zinc is especially important in such hydrolytic enzymes (cf. Ref. 94). For example, the enzyme carboxypeptidase A uses zinc in a typical bifunctional role, at the same time activating a carbonyl for addition by coordinating with its oxygen and activating a water molecule to act as a nucleophile [95–100]. We produced a model for this type of process by using metal complexing as the substrate binding force and a coordinated oxime as the nucleophile (**17**) [101]. The geometry of this compound means that the Lewis acidic zinc and the basic oxime anion can co-exist without quenching each other; the electrons can flow from one to the other only through the bridging carbonyl group of the substrate (**18**). Consequently, the anion of **17** reacted with metal-bound substrate **19** to transfer the acetyl group to the oxime anionic oxygen, and then the intermediate **20** rapidly hydrolyzed. In this process the metal ion is serving multiple functions; it binds the substrate, acidifies the oxime, coordinates to the carbonyl oxygen of the transferring acetyl group, and then catalyzes hydrolysis of intermediate **20**. It was our first example of such an enzyme-like process, and we built on it further.

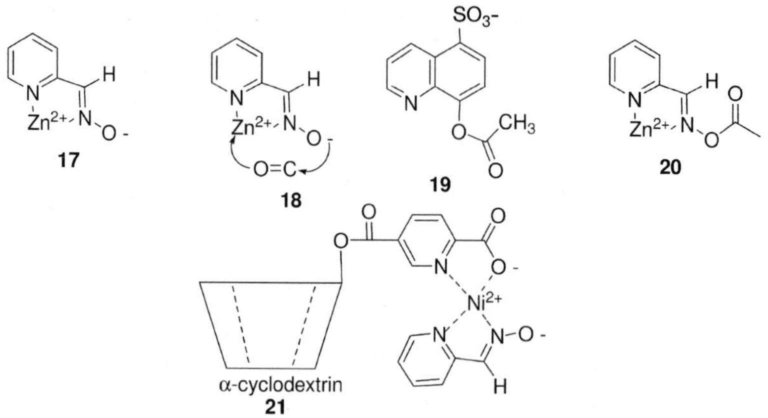

Catalyst **17** is effective only with substrates that can bind to the metal ion, so we attached it – coordinated as its Ni^{2+} derivative – to the secondary face of α-cyclodextrin in catalyst **21** [102]. This was then able to use the metallo-oxime catalysis of our previous study, but with substrates that are not metal ligands, simply those that bind hydrophobically into the cyclodextrin cavity. As hoped, we saw a significant rate increase in the hydrolysis of *p*-nitrophenyl acetate, well beyond that for hydrolysis without the catalyst or for simple acetyl transfer to the cyclodextrin itself. Since there was full catalytic turnover, we called compound **21** an "artificial enzyme" – apparently the first use of this term in the literature. The mechanism is related to that proposed earlier for the enzyme alkaline phosphatase [103].

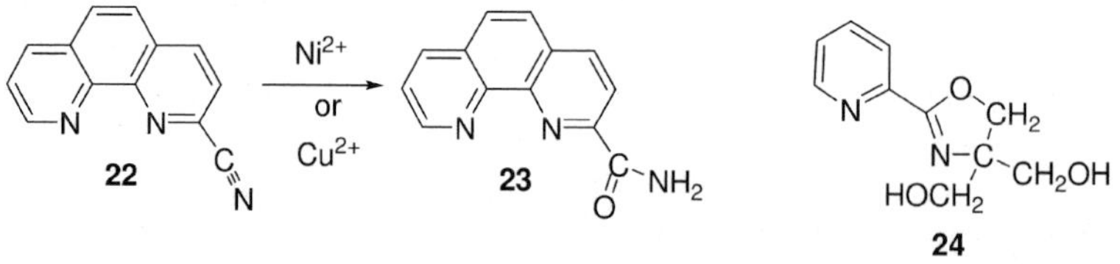

Nitriles can be hydrated enzymatically to form amides. In a model system we showed [104] that **22** can be converted into the amide **23** when metal ions are coordinated into the phenanthroline system. With Ni^{2+} the rate acceleration was 10^7, while with Cu^{2+} the hydration was accelerated by 10^9. These are huge rate increases. Much of the driving force is related to the fact that the cyano group is not itself a strong metal ligand in **22**, but the transition state for the hydration is metal coordinated. Also, we used a metal ion to organize the intracomplex reaction of a ligand–ligand reaction [105] in which tris-hydroxymethylaminomethane (Tris) adds to 2-cyanopyridine to form the adduct **24**. Again, the rate was very large and, more to the point, the addition of Tris occurs even though the concentration of water is 10^4 times that of Tris. In the absence of Tris the metal ion simply catalyzed the hydration of 2-cyanopyridine to its amide, as with **22** above. The coordination of both reactants to the same metal ion has produced selectivity that was induced by the coordination.

In models for carboxypeptidase A we showed the intracomplex catalyzed hydrolysis of an ester by a metal ion and a carboxylate ion [106], which are the catalytic groups of carboxypeptidase A. Some mechanistic proposals for the action of carboxypeptidase involve an anhydride intermediate that then hydrolyzes to the product and the regenerated enzyme. Although we later found convincing evidence that the enzyme does not use the anhydride mechanism in cleaving peptides [96–99], it may well use such a mechanism with esters. In a mimic of part of this mechanism we showed [107], but see also Ref. 108, that we could achieve very rapid hydrolysis of an anhydride by bound Zn^{2+}, which is the metal ion in the enzyme. In another model, a carboxylate ion and a phenolic hydroxyl group, which are in the enzyme active site, could cooperatively catalyze the cleavage of an amide by the anhydride mechanism [109].

Catalysis by Zn^{2+} is ambiguous, since it is not clear whether the metal ion is coordinated to the carbonyl group of the substrate, as it is in the enzyme. Thus, we examined the cleavage of an amide by a combination of Co^{3+} and a carboxylate group. Since Co(III) is "substitution inert," we prepared a complex in which it was directly coordinated to the carbonyl oxygen of a substrate amide, and in which a neighboring carboxylate ion or phenol group was a potential second catalytic function [110]. Indeed, the phenol group was able to assist the hydrolysis by protonating the leaving amino group, but the carboxylate ion was not effective. The mechanism is intellectually related to that used by the enzyme [97, 98], but the details differ.

By contrast with the absence of catalysis by an internally attached carboxylate ion in the above study, an external carboxylate species and, even more, an external phosphate species were catalysts with the cobalt complexed amide. Thus we examined a process like that just described in which a *phosphonate* group was internally attached to the cobalt complex [111]. In this case it did act as a sequential base/acid catalyst, as the carboxylate ion does in the enzyme, i.e., it first delivered a hydroxide group to the complexed carbonyl, acting as a base, and then the proton that it had accepted was delivered to the leaving group, with the phosphonic acid acting as a general acid. This sequence is indeed like that in the enzyme. We suggested that the phosphonate is more effective because it has a higher pK_a than the attached carboxylate, and in the enzyme the carboxylate has an abnormally high pK_a. The phosphonate is a better model for the enzyme carboxylate in the artificial enzyme.

These cobalt systems are useful models of enzymatic mechanisms, but they are not turnover catalysts as enzymes are. To achieve turnover we constructed ligand **25** in which a metal-coordinating group links two cyclodextrin rings. As its metal complex it was a good catalyst for the hydrolysis of substrates **26** and **27** that could bind into both cyclodextrins and stretch across the bound metal ion. (Later we will describe the same principle applied to selective oxidation reactions.) Ligand **25** as its Cu^{2+} complex gave as much as a 10^5-fold rate acceleration in the ester hydrolysis [112, 113]. With an added nucleophile that also binds to the Cu^{2+} ion, the reaction is accelerated by over 10^7. The mechanism deduced (**28**) – in which the metal ion acts as a Lewis acid by coordination to the substrate carbonyl and also delivers a bound hydroxide ion to the ester carbonyl

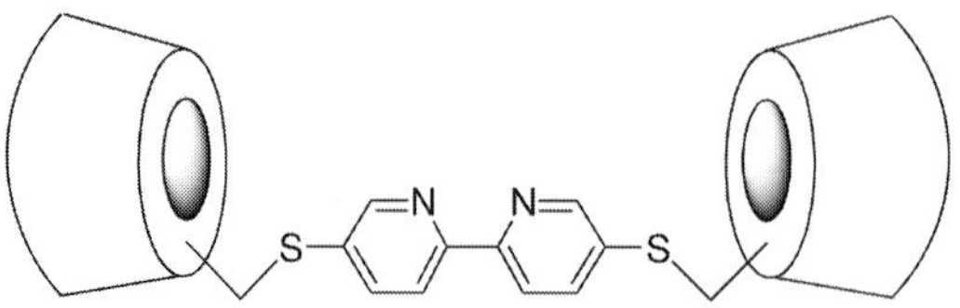

25

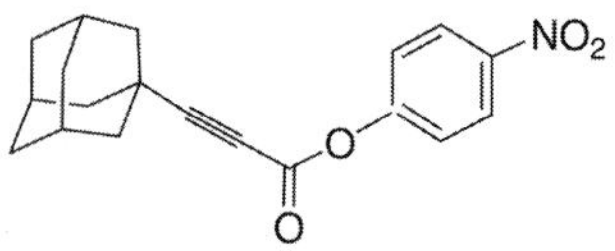

26

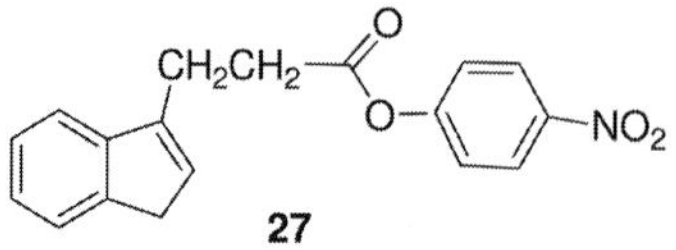

27

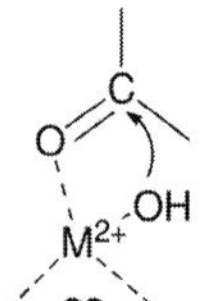

28

group – is like both that in the enzyme carbonic anhydrase and the one we proposed for histone deacetylase [114, 115]. In those enzymes Zn^{2+} is the catalytic metal ion.

We have pursued such ester hydrolysis by artificial enzymes further. With a cyclodextrin dimer related to **25** we have hydrolyzed an ordinary doubly bound ester, not just the more reactive nitrophenyl esters [116], with catalytic turnovers. Also, with a catalyst consisting of a cyclodextrin linked to a metal ligand carrying a Zn^{2+} and its bound oxime anion, we saw good catalyzed hydrolysis of bound phenyl esters with what is called burst kinetics (fast acylation, slower deacylation), as is seen with many enzymes [117].

Artificial enzymes with metal ions can also hydrolyze phosphate esters (alkaline phosphatase is such a natural zinc enzyme). We examined the hydrolysis of *p*-nitrophenyl,diphenylphosphate (**29**) by zinc complex **30**, and also saw that in a micelle the related complex **31** was an even more effective catalyst [118]. Again the most likely mechanism is the bifunctional Zn–OH acting as both a Lewis acid and a hydroxide nucleophile, as in many zinc enzymes. By attaching the zinc complex **30** to one or two cyclodextrins, we saw even better catalysis with these full enzyme mimics [119]. A catalyst based on **25** – in which a bound La^{3+} cooperates with H_2O_2, not water – accelerates the cleavage of bis-*p*-nitrophenyl phosphate by over 10^8-fold relative to uncatalyzed hydrolysis [120]. This is an enormous acceleration.

Phosphate ester cleavage can also be achieved with artificial enzymes using both a metal ion and an additional catalytic group, as in the amide and ester hydrolyses described above. In our first example, catalysts **32** and **33** combined a Zn^{2+} with a thiophenol and an imidazole group respectively [121]. The rigid structure prevented the

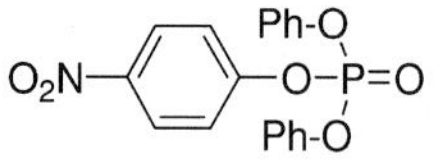

29

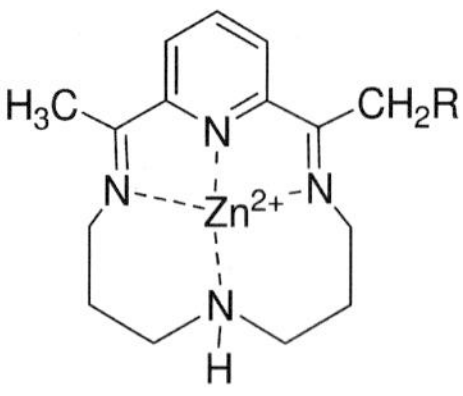

30, R = H
31, R = n-C$_{16}$H$_{33}$

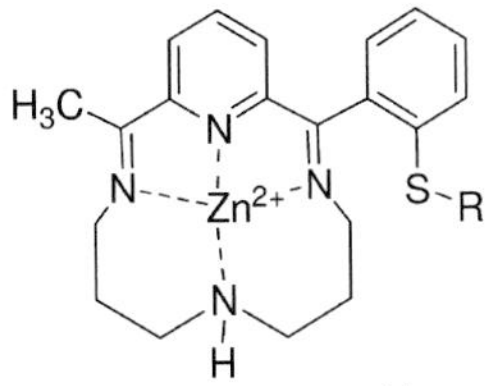

32, R = H
33, R =

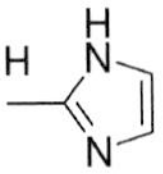

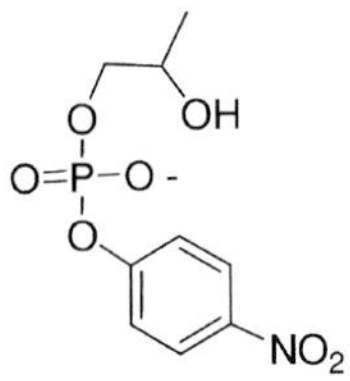

34

imidazole or SH coordinating to the metal ion. In electrical terms, they can feed electrons to the metal ion only when there is bridging by a substrate group; the resulting electron flow leads to catalysis. The bifunctional catalysis led to cyclization of substrate **34**.

As another example, we prepared disubstituted cyclodextrin **35** in which one substituent was a metal-binding tren group while the other was an imidazole [122]. Zn^{2+} complexed to the tren group gave good rate acceleration in the hydrolysis of bound catechol cyclic phosphate **36**, which was fastest when the two catalytic groups were attached to opposite sides of the cyclodextrin so they could not bind each other. The geometry of the complex led to the selective formation of product **37** rather than **38**; both are formed equally by ordinary hydrolysis without the catalyst.

The next section describes further mimics of ribonuclease enzymes. However, we mention here one additional study on the cleavage of ribonucleotides in which metal ions were used [123]. Cyclization of uridyluridine **39**, and hydrolysis of the resulting cyclic nucleotides **40**, was catalyzed by Eu^{3+} much more effectively than by Zn^{2+}, and some added ligands increased the rates.

1.3.3
Artificial Ribonucleases

Ribonuclease A is a member of a group of enzymes that cleave RNA using general acid–base catalysis without a metal ion in the enzyme. In ribonuclease A, such catalysis is performed by two imidazoles of histidine units, one as the free base (Im) and the other, protonated, as the acid (ImH⁺). To mimic this in an artificial enzyme, we prepared β-cyclodextrin bis-imidazoles **41** [124]. The first one was a mixture of the

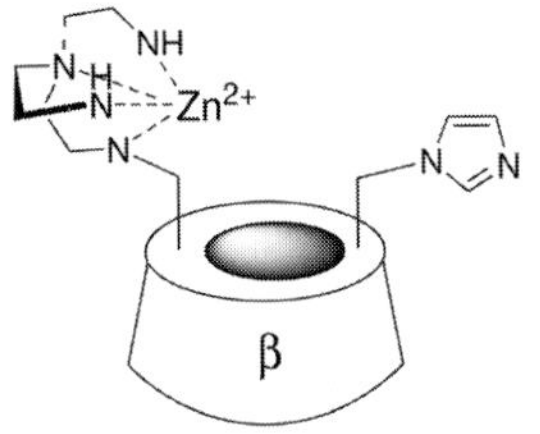

35

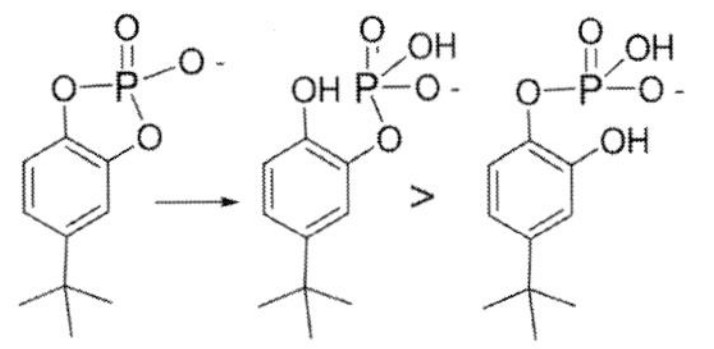

36

37

38

39 → **40** + uridine

6A,6C and 6A,6D isomers (the seven glucose units of the cyclodextrin are labeled A through G). (As described below, we were later able to prepare as pure catalysts all the isomers of the bis-imidazole cyclodextrin with the imidazole rings on the primary carbons of the ring. The geometric dependence of catalysis indicated the mechanism involved.)

On examining this catalyst mixture in the hydrolysis of the cyclic phosphate of 4-*t*-butylcatechol **36**, we saw a bell-shaped pH rate profile, indicating bifunctional acid–base catalysis of the hydrolysis as in the enzyme ribonuclease. A catalyst carrying only one imidazole showed only base catalysis, by the unprotonated imidazole group Im. Thus, in catalyst mixture **41**, one imidazole was acting as a base – delivering a water molecule to the phosphate group of the bound substrate – while the imidazolium ion of the other catalytic group played a role as a general acid. At the time we thought that this imidazolium ion might be simply protonating the leaving group of the phosphate, as was normally assumed for the enzyme ribonuclease A, but our later work revealed a more interesting role.

Simple hydrolysis of substrate **36** with base in solution furnishes an essentially equal mixture of **37** and **38**, since the *t*-butyl group is too remote to have any influence. However, catalysis by the artificial enzyme mixture **41** gave **37** only. The geometry of the catalyst–substrate complex directs the attack by water (hydrogen bonded to the imidazole) in a line perpendicular to the ring axis (cf. **42**). When we moved the two imidazole groups out further, in catalyst **43**, they were then able to deliver the water in a

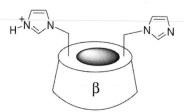

41, a mixture of the A,C and A,D positional isomers, as the monoprotonated species.

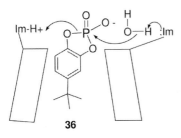

36

42, a possible catalytic
bifunctional mechanism
for cleavage of **36** by
catalyst **41**

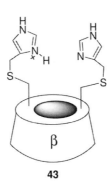

β

43

different direction, to afford essentially only **38** [125]. Molecular models were consistent with this change in preferred geometry of attack on the cyclic phosphate.

We developed an assay for the cleavage of uridyluridine (**39**, UpU) by various catalysts, and used it to study the cleavage of this dimeric piece of RNA [126]. We saw that high concentrations of imidazole buffer could catalyze this cleavage, mimicking the high effective local concentrations of imidazole in the enzyme, and concluded that with this buffer there was sequential base, then acid, catalysis [127]. Of course, simultaneous catalysis by two different buffer species by a three-body collision is unlikely unless they are linked in the same catalyst – the enzyme or the artificial enzyme.

Further work on this simple imidazole catalysis showed that, as well as cyclization of UpU to a cyclic phosphate, 3,5-linked UpU (**39**) also isomerized to the 2,5-linked isomer **44** [128]. Such an isomerization requires a phosphorane intermediate (**45**) and its pseudo-rotation so that the C-3 oxygen can become the leaving group. Our studies showed that in this buffer-catalyzed reaction the ImH$^+$ catalyst was not simply protonating the leaving group – it was first protonating the phosphate anion of the UpU so that a phosphorane intermediate could be formed in both the cleavage and isomerization. We suggested that this could also be the mechanism in the enzyme ribonuclease itself, but this is still a matter of controversy. In any case, we soon found evidence that this mechanism – proceeding through a five-coordinate phosphorus, a phosphorane –operated with our artificial enzyme symbolized by **41**.

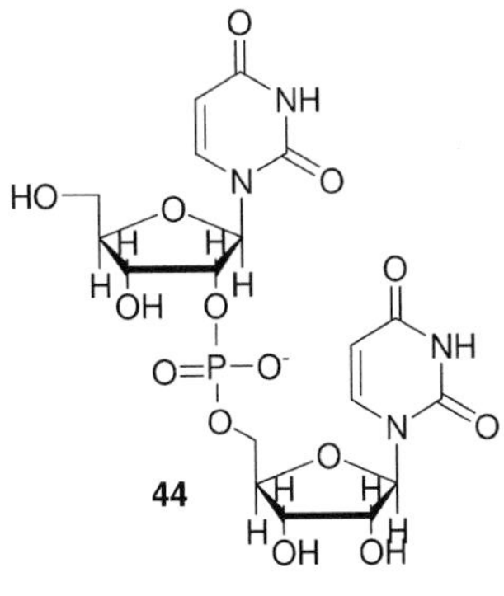

44

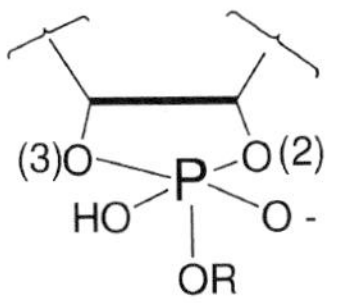

45, a phosphorane
intermediate with an apical
O-2 and an equatorial O-3.
To form **44** it must pseudorotate
so the O-3 becomes apical.

46, the A,B isomer   **36**   **47**   **37**

As mentioned above, we first made the cyclodextrin bis-imidazole catalyst with the imidazoles attached to primary carbons that were as far apart as possible (A,D glucose residues) or almost that far (A,C residues). We then made all three isomers selectively – A,D and A,C and A,B – to see how their geometry affected the catalyzed hydrolysis of substrate **36** [129]. All three showed the bell-shaped pH vs. rate curves, indicating that there was both base and acid catalysis, by an Im and an ImH⁺ group. They all catalyzed hydrolysis of the substrate but, remarkably, the best catalyst by far was the A,B isomer **46**, i.e., that with the acid and base groups right next to each other. This was completely inconsistent with a mechanism (**42**) in which a water molecule (bound to an Im) attacks the phosphate while the leaving group departs, assisted by the ImH⁺. Such an in-line mechanism would require the Im and ImH⁺ to be 180° apart, more or less, and so the A,D isomer should have been the best catalyst. Clearly, the ImH⁺ was playing a different role, protonating the phosphate oxyanion to facilitate formation of a phosphorane intermediate (**47**). This is the mechanism we had deduced for simple Im/ImH⁺ buffer-catalyzed hydrolysis of UpU (vide supra).

By proton inventory, a technique that determines whether acid and base groups act simultaneously, we found that hydrolysis of **36** by artificial enzyme **44** involves two protons moving in the transition state [130]. Thus, ImH⁺ of **46** is hydrogen bonded to a phosphate oxyanion of bound substrate **36**; water hydrogen bonded to the Im then attacks the phosphorus, and as the O–P bond forms the ImH⁺ proton transfers (along with the water proton) to produce the phosphorane monoanion **47**. This then goes on to the cleaved product in later catalyzed steps before there is time for pseudo-rotation. These general conclusions have been described and summarized in several publications [131–137].

1.3.4
Artificial Enolases and Aldolases

We examined the ability of our bis-imidazole cyclodextrin artificial enzymes to perform other bifunctionally-catalyzed reactions, where again the availability of the A,B and A,C and A.D isomers let us learn mechanistic details. As an important example, we examined three isomeric catalysts' ability to promote the enolization of substrate **48**, which binds into the cyclodextrin cavity in water [138]. Here there was again a strong preference among the isomers, but it was the A,D isomer **49** that was the effective catalyst! It was also more effective than a cyclodextrin mono-imidazole that cannot use the bifunctional mechanism.

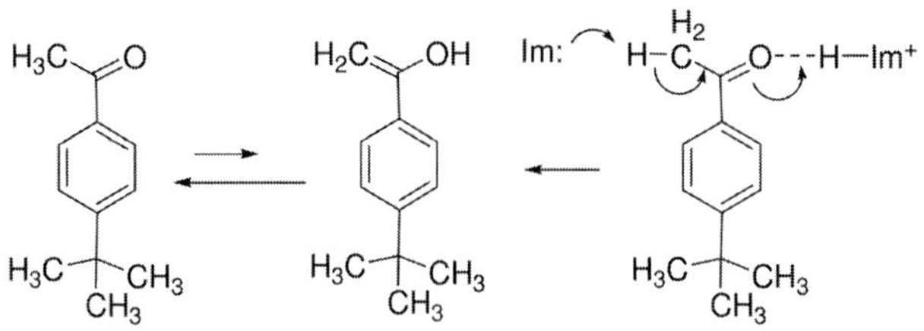

48   **50**

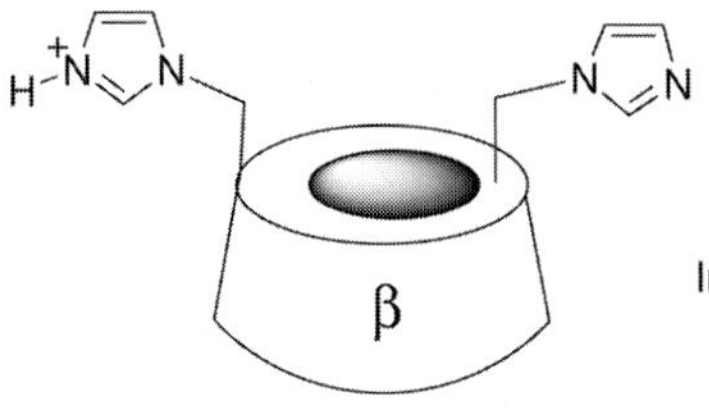

49, the A,D isomer 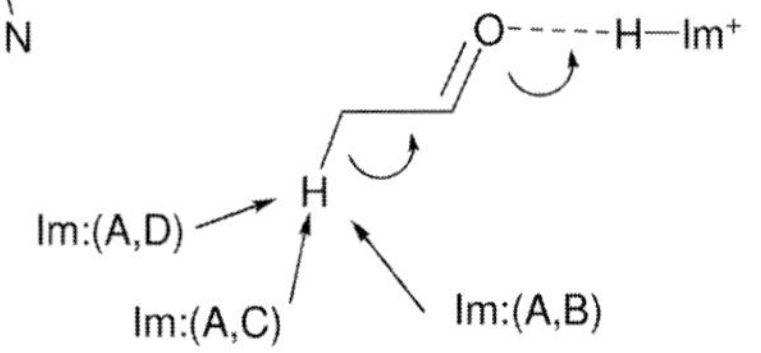   **51**, the A,D is preferred

Cyclodextrin bis-imidazole catalyzes enolization by a bifunctional mechanism in which the ImH$^+$ is hydrogen-bonded to the carbonyl oxygen while the Im removes the neighboring methyl proton (cf. **50**). As expected from this, there was a bell-shaped pH vs. rate profile for the process. In the transition state two protons will move simultaneously, as in the hydrolysis reaction described above. Thus we indeed have a powerful tool to determine the geometric requirements for simultaneous bifunctional catalysis, a tool that could be of quite general use.

With enolization, we were able to understand the preference for the A,D isomer **49** in stereoelectronic terms. Models show that all three isomers can achieve geometries in which the ImH$^+$ can hydrogen bond to the carbonyl oxygen while the Im can reach the methyl proton, but the direction of attack on that proton differs among the isomers. The preferred isomer, the A,D species, removes the proton by a non-linear attack (cf. **51**), pushing the electrons toward the carbonyl group. This is presumably true for all enolizations, although techniques have not existed before to determine it.

Enolization can be part of an aldol condensation. We examined the aldol cyclization of compound **52** to **53** catalyzed by the bis-imidazole cyclodextrin artificial enzymes, and again saw that the A,D isomer was the preferred catalyst [139]. This was not an obvious result; the rate-limiting step in this case is cyclization of the enol, which is

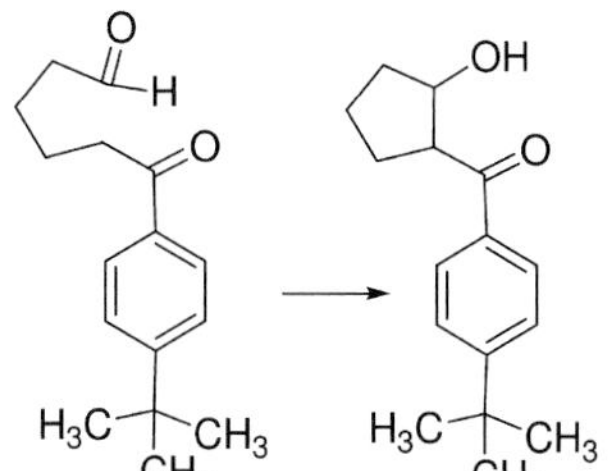

52   **53**

reversibly formed. In the absence of the catalyst, enolization – indicated by deuterium exchange – occurred rapidly next to the more reactive aldehyde group, as expected. However, with catalyst **49** this was reversed, and most of the deuterium exchange occurred next to the ketone group, in reach of the catalytic imidazoles. This enol cyclized with the aldehyde in the slow step of the sequence, but again this step had a preference for the A,D isomer **49** of the catalyst.

The situation is complex. In another study we examined the cyclization of compound **54** catalyzed by cyclodextrin bis-imidazoles [140]. This dialdehyde can perform the intramolecular aldol reaction using the enol of either aldehyde to add to the other aldehyde, forming either **55** or **56**. In solution with simple buffer catalysis both compounds are formed almost randomly, but with the A,B isomer **46** of the bis-imidazole cyclodextrin there was a 97% preference for product **56**. This is consistent with the previous findings that the catalyst promotes enolization near the bound phenyl ring, but in this case the cyclization is most selective with the A,B isomer **46**, not the A,D that we saw previously. Again the enolization is reversible, and the selectivity reflects the addition of an enol to an aldehyde group. The predominant product is a mixture of two stereoisomers, **56A** and **56B**. Both were formed, and were racemic despite the chirality of the cyclodextrin ring.

We have also examined the use of cyclodextrin-derived artificial enzymes in promoting *bimolecular* aldol reactions, specifically those of *m*-nitrobenzaldehyde (**57**) and of *p*-*t*-butylbenzaldehyde (**58**) with acetone [141]. Here, we examined a group of mono-substituted cyclodextrins as catalysts (e.g. **59**), as well as two disubstituted β-cyclodextrins (e.g. **60**) (10 catalysts in all). They all bound the aldehyde components in the cyclodextrin cavity and used amino groups of the substituents to convert the acetone into its enamine. An intracomplex reaction with **58** and hydrolysis of the enamine product then afforded hydroxyketone **61** (cf. **62**). These catalysts imitate natural enzymes classified as Class I aldolases.

Although *m*-nitrobenzaldehyde (**57**) is well bound into the cyclodextrin cavity of our above catalysts, there was essentially no catalysis of its reaction with acetone. The aldehyde group is too inaccessible in the complex. However, with **58** there was good catalysis by a cyclodextrin carrying only one ethylenediamine group on its secondary face, and also by cyclodextrins with two groups on the primary A and B methylenes (e.g. **60**), with imidazoles as base/acid groups and a primary amine to form the acetone

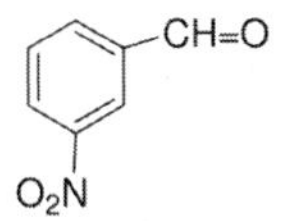

57 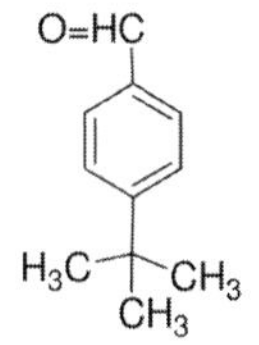 **58** 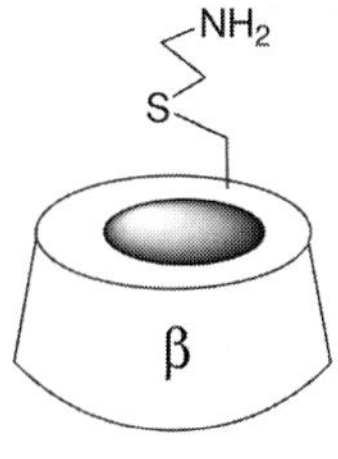 **59** 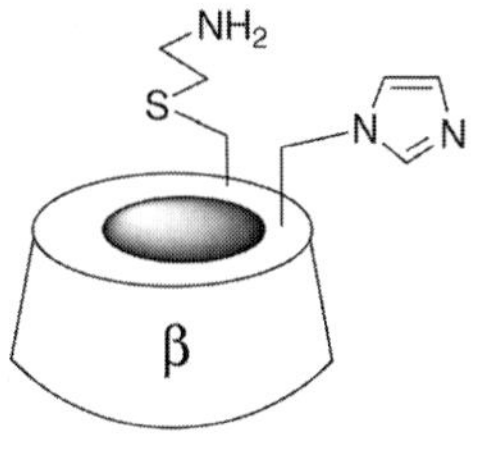 **60**, the A,B isomer

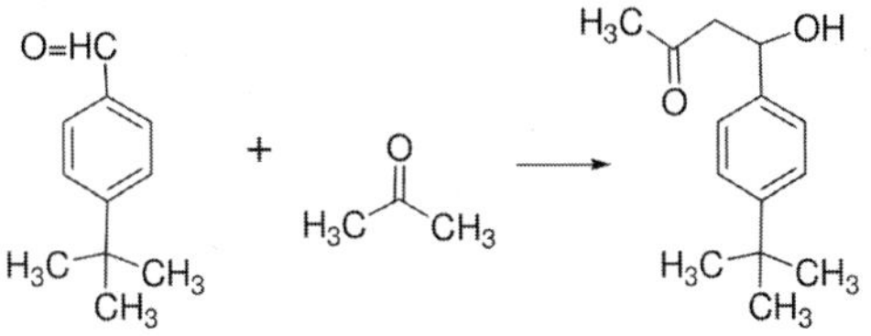

58 + → **61** 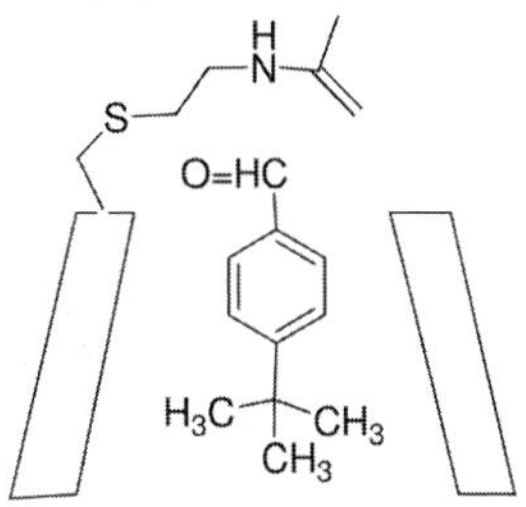 **62**

enamine. Some of the artificial enzymes also catalyzed the condensation of **58** with cyclopentanone. In this series we are using the catalytic groups on the cyclodextrin not just to perform acid–base processes but also to bind an external ketone as its enamine and thus promote intracomplex reactions.

1.4
Cytochrome P-450 Mimics

Enzymes exhibit wonderful selectivities, directed by the geometries of the enzyme–substrate complexes. For many practical purposes, learning how to imitate such selectivities is even more important than achieving rate accelerations at the enzyme level. Of course, selectivity is a reflection of relative rates, but a selective rate advantage of only one-hundred-fold could be enough to produce one product rather than either another or a mixture. This is easier to achieve than the huge overall rate accelerations that most enzymes exhibit.

1.4.1
Aromatic Substitution in Cyclodextrin Complexes

In one of our earliest approaches to such biomimetic selectivity, directed by the geometry of the catalyst–substrate complex, we examined the directed substitution of an aromatic ring bound into the cyclodextrin cavity [142, 143]. Anisole was chlorinated by HOCl entirely in the para position when it was bound into α-cyclodextrin, while in

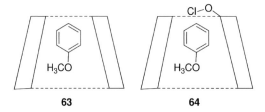

63   **64**

water solution without the cyclodextrin the product was 60% *p*- and 40% *o*-chloroanisole. The simplest idea could be that the cyclodextrin simply blocks the ortho positions in complex **63**, but this is not correct. The cyclodextrin actually accelerates the reaction of the para position, so it is a catalyst, not just a blocker.

Also, in water solution the chlorination of anisole by HOCl shows second-order kinetics in HOCl; the chlorinating agent is the more powerful Cl_2O, which is in equilibrium with two HOCl molecules [143]. However, when the cyclodextrin is present the chlorination shows only first-order dependence on [HOCl], indicating that a different chlorinating species is involved in the catalytic reaction. We also examined the reaction using β-cyclodextrin instead of α-cyclodextrin. Again, there was no ortho chlorination of anisole in the complex, but now there was no acceleration of the para chlorination. Thus, the catalyzed para chlorination process is faster with the well-fitting α-cyclodextrin than with the more loosely fitting β-cyclodextrin.

From all this we concluded that the catalysis by cyclodextrin involves the reversible formation of a cyclodextrin hypochlorite (**64**), which can transfer chlorine to the anisole para position but cannot reach the ortho position. The reaction is faster when the RO–Cl group is held closer and more rigidly in the smaller α-cyclodextrin.

It is interesting that this artificial enzyme induces such selectivity whereas the enzyme chlorinase does not. The real enzyme produces a 60:40 mixture of para and ortho chloroanisoles, presumably because it generates HOCl that reacts with free anisole in solution. The artificial enzyme is more like a typical enzyme – which would be highly selective in its products – than is this particular natural chlorinating enzyme!

Further insight came from our study of other aromatic substitution reactions. When we blocked the para position of anisole in compound **65**, we saw that ortho chlorination was blocked by binding with α-cyclodextrin, so the only reaction was from the substrate that was in free solution, not that which was bound. However, with *p*-cresol (**66**) there was still, of course, ortho chlorination but now it was *catalyzed* by the α-cyclodextrin. When *p*-cresol binds to the cyclodextrin, the polar phenol or phenoxide group will be out of the cavity, bringing the ortho positions within reach of the cyclo-

65   **66**

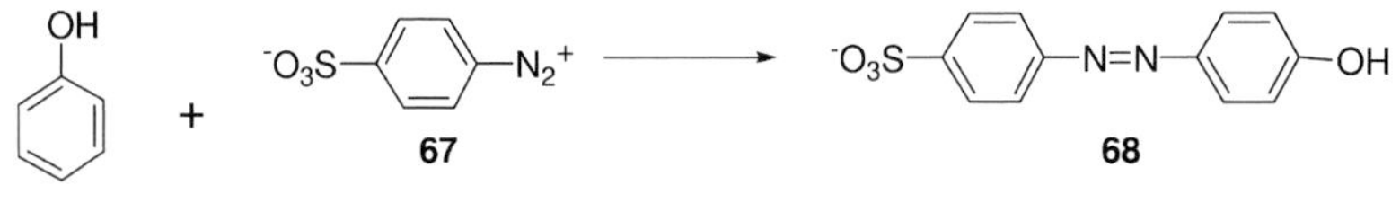

dextrin hydroxyl groups. Therefore, in the selective reaction either para or ortho substitution can be catalyzed if the geometry of the complex is suitable.

One important control was the study of the diazo coupling of phenol with **67**. Here para coupling formed **68**, but the reaction was inhibited by α-cyclodextrin. The diazonium group of **67** cannot be delivered by binding to a cyclodextrin hydroxyl group, in contrast to a chlorine atom. This supports the idea that cyclodextrin catalysis of chlorinations involves binding of both the chlorine atom and substrate, holding them together for rapid selective reaction.

In a further study, α-cyclodextrin that was partially methylated, blocking the OH groups at C-2 and C-6 and leaving the C-3 OH group free [144], was found to catalyze the para chlorination of anisole. Therefore, at least the cyclodextrin C-3 OH can be the catalytic group. Furthermore, the binding of anisole was stronger than that in α-cyclodextrin, probably reflecting the flexible capping of the cavity described earlier [84]. Thus it was possible to achieve 99% para chlorination of anisole (without the flexible capping, there was always some free anisole in solution that led to some ortho chlorination as well). We also synthesized a polymer, consisting of linked α-cyclodextrins, that could be used in a flow mode to produce anisole that was over 99% para chlorinated, because there was essentially no free anisole in equilibrium with the polymeric α-cyclodextrin and the reaction within the complex was completely selective [144].

1.4.2
Selective Photochemical Reactions

Our principal target in P-450 mimics was the selective oxidation of saturated carbons directed by geometric control, not by intrinsic reactivity. In our first study, we examined selective geometrically controlled attack on aliphatic C–H bonds by photo-excited benzophenones [145]. In a process we labeled "remote oxidation", photolysis of a long-chain ester **69** of benzophenone-4-carboxylic acid afforded insertion into CH_2 groups far into the chain.

In such an insertion the benzophenone is excited to its triplet state, resembling a C–O diradical. The oxygen atom of the ketone then removes a hydrogen atom from the methylene group, leading to a carbinyl radical and the hydroxy diphenylmethyl radical formed by hydrogen atom addition to the benzophenone. This pair of radicals couples

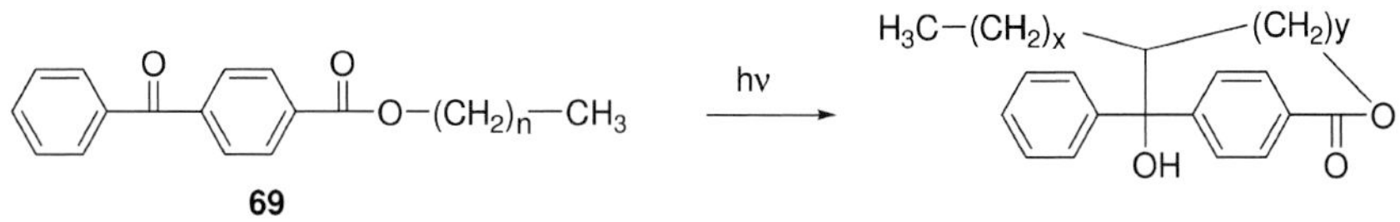

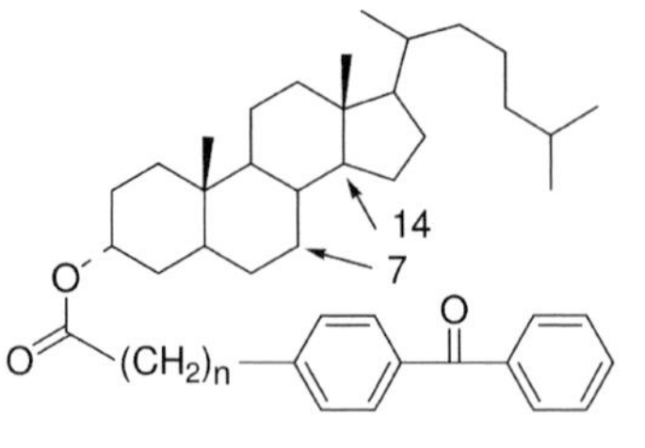

70, n = 2

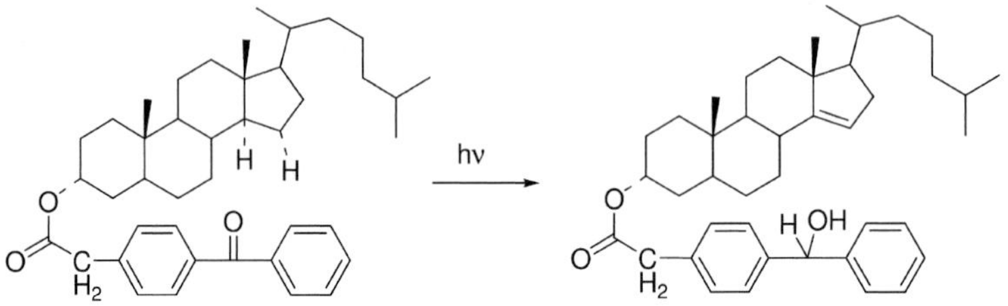

71   **72**

to make the new carbon–carbon bond. There was no great selectivity (with a C_{20} chain there was appreciable insertion into carbons 11–16) but the method gave an indication of the accessible conformations of the molecule [146]. We later used this method to determine the conformations of the flexible chains in membranes [147, 148], and others have adapted it for photo-affinity labeling in proteins.

With inflexible substrates, this photochemical functionalization process showed significant selectivity. As our first example, photolysis of steroid ester **70** led to carbonyl insertion into the C–H bonds at C-7 and C-14 [149, 150]. Other steroid esters afforded selective functionalizations in various positions [e.g. 151, 152], and compound **71** directly dehydrogenated the steroid to form the product olefin **72** [150].

Later work gave additional examples, and we were able to elucidate the process by which **72** is formed [153]. After initial removal of the C-14 hydrogen by the oxygen of the benzophenone triplet, the resulting C-14 radical inverts to bring C-15 near the reagent radical; the C-15α hydrogen is then transferred to produce product **72**. This was demonstrated by use of C-15α deuterium-labeled steroid. Apparently, in the more rigid **71** such hydrogen transfer in the radical pair is easier than is coupling to form a carbon–carbon bond, whose cyclic product would be strained.

We also saw directed functionalizations when a benzophenone reagent (**73**) was not directly attached to a steroid or long alkyl chain substrate (**74**), but was linked by hydrogen bonding [151].

In all the above studies the benzophenone units were singly attached (or with the previous example, complexed) to the substrates, so that there was intramolecular or

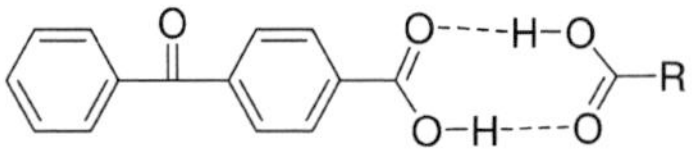

73   **74**

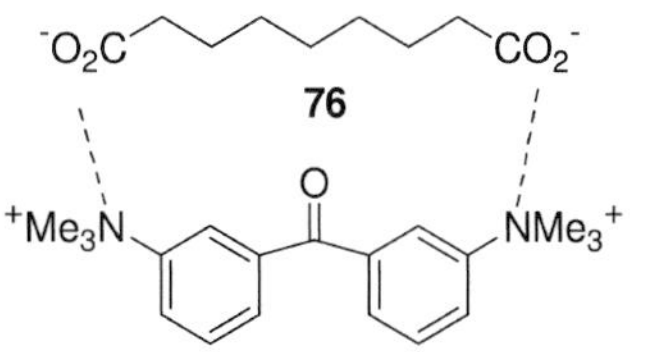

intracomplex functionalization directed by the conformations of freely rotating units. To overcome this flexibility, we studied the functionalization of dicarboxylic acids doubly complexed with disubstituted benzophenones [154]. Two sorts of complexing were used: benzophenone dication **75** ion-paired with substrate dicarboxylate ion **76**, and benzophenone dicarboxylate **77** hydrogen bonded to substrate dicarboxylic acid **78**. We saw high selectivities for attack at the two central (equivalent) carbons with a substrate of the correct length to fit well across the benzophenone, while there was less selectivity with a longer substrate with flexible conformations in the complexes. As described later, such double complexing of substrates was of general interest and general utility in other artificial enzymes.

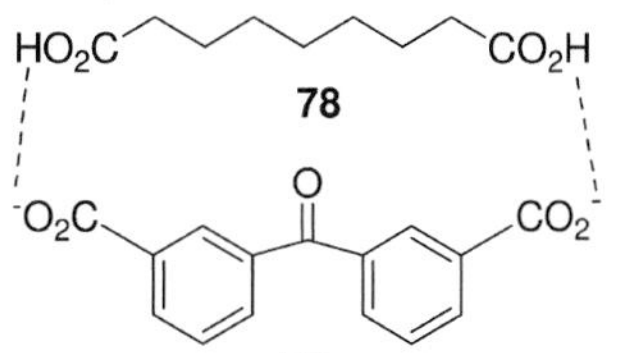

1.4.3
Directed Halogenations

Irradiation of the benzophenone-linked steroids in the presence of CCl_4 solvent gave some chlorination of the steroids. Also, irradiation in the presence of $BrCCl_3$ afforded no benzophenone products, only bromosteroids from a free radical chain process initiated by the irradiation [155]. In the presence of $PhICl_2$ we saw only chlorosteroids, again by a radical chain process [155]. In both cases the rather selective hydrogen abstraction led to exclusive substitution of tertiary hydrogens of the steroids, in particular at C-9 and C-14. C-9 bromo- and chlorosteroids **79** and **80** could be dehydrohalogenated to form the 9(11) olefin **81**. This afforded an easy entry into corticosteroids, which have an oxygen atom at C-11 and often, usefully, have a fluorine at C-9.

Such normal free-radical substitution processes might be geometrically directed to otherwise unremarkable positions in substrates, as the benzophenone reactions had been. Consequently, in our first example, we found that intramolecular chlorination could be directed by attachment of a $PhICl_2$ group to the steroid [156], i.e., compound **82** directed chlorination exclusively to C-14 (**83**), while **84** directed halogenation exclusively to C-9 (**85**). As shown in Scheme 1.1, an intramolecular hydrogen abstraction by

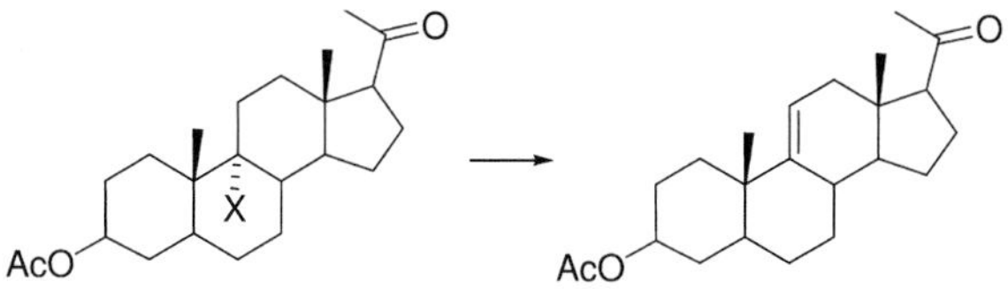

79, X = Br
80, X = Cl

81

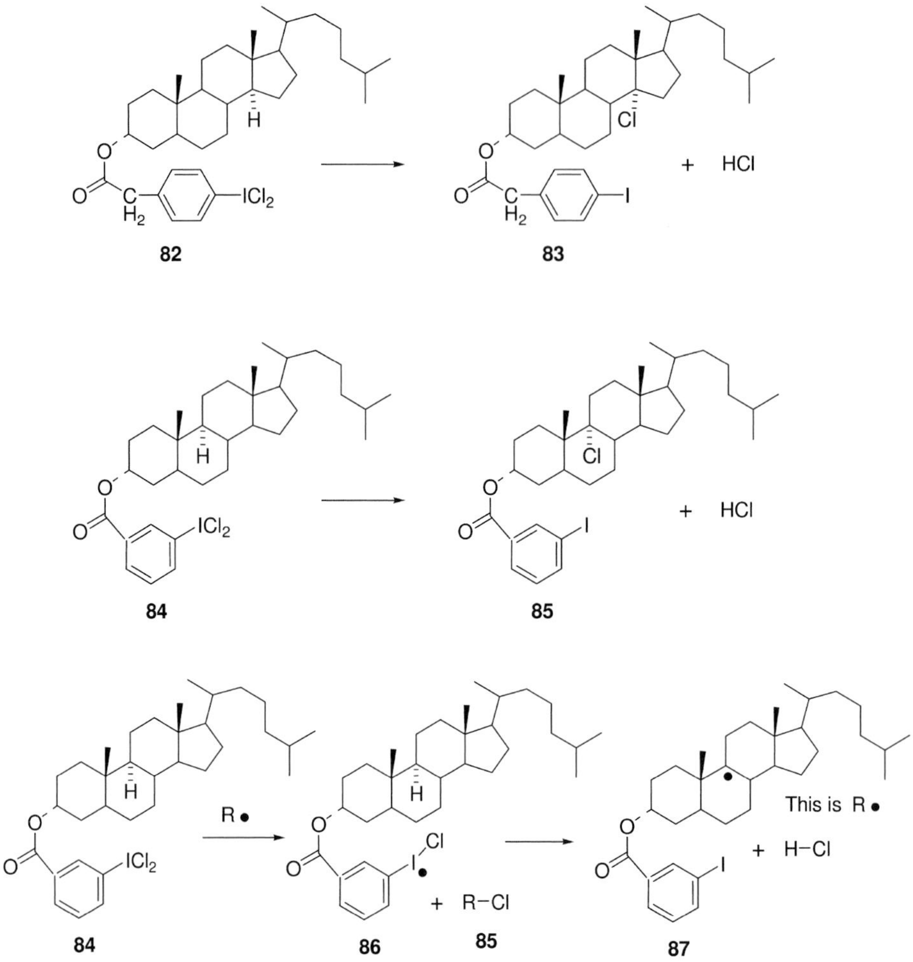

Scheme 1.1

an attached PhICl radical (**86**) to generate (**87**) is followed by an intermolecular chlorine transfer from another substrate to generate its attached radical.

However, because there is only a stoichiometric amount of the chlorinating agent, side reactions mean that some substrate is left unchlorinated. Also, the attached PhICl$_2$ reagent must be made by adding Cl$_2$ to the easily produced PhI ester. To better achieve the same results, using chlorine transfer from an external radical to generate the attached PhICl radical, we employed a process we called a Radical Relay Mechanism [157] (Scheme 1.2). Here an external PhICl radical transfers its chlorine atom to the iodophenyl ester group attached to the substrate, and then the intramolecular hydrogen atom abstraction occurs. Thus we can use a slight excess of the PhICl$_2$ reagent to complete the process, and there is no need to add Cl$_2$ to the attached iodophenyl group, which acts as a template to steer the chlorinations.

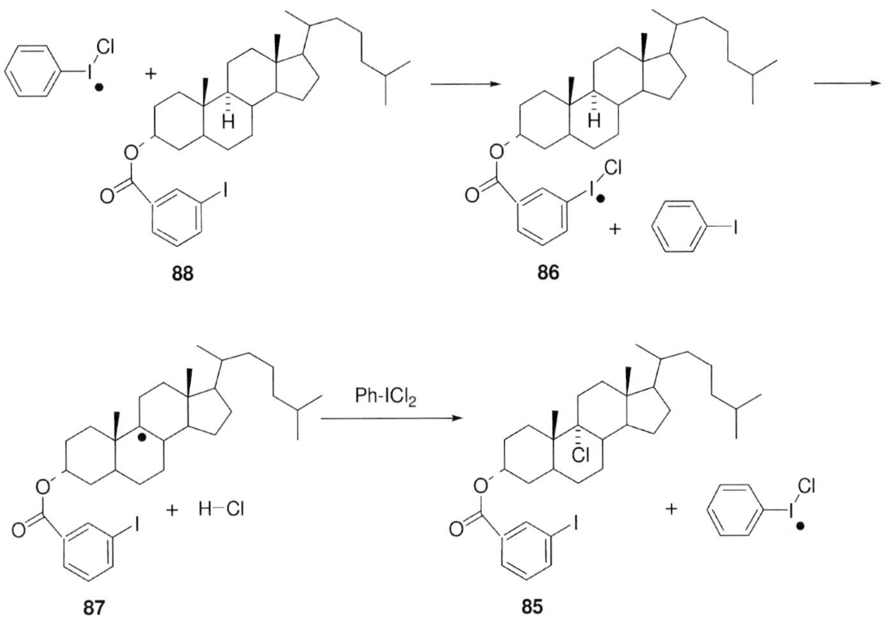

Scheme 1.2

As expected, compound **89** underwent C-14 chlorination and **90** underwent C-17 chlorination. Other chlorinating agents could also be used. In particular, SO_2Cl_2 was able to selectively chlorinate **88–90** by transferring a Cl atom from a $ClSO_2$ intermediate radical (liberating SO_2), and then generating this radical after Cl transfer to the steroid radical.

Why does this work? Why does an externally generated reagent not directly attack the substrate but instead transfer an atom to the template, which then leads to attack on the C–H bond with geometric control of the attack position? The reason is the same as that which causes many hydrolytic enzymes to cleave a peptide bond in two steps by first forming an ester intermediate with a serine hydroxyl group and then later using water to hydrolyze that ester. By the time the water attacks, with loss of translational entropy, the amine component of the original peptide bond is freely moving and has gained translational entropy. In our radical relay mechanism, the PhI molecule – formed by chlorine atom transfer to the large accessible iodine atom of the template – is freely

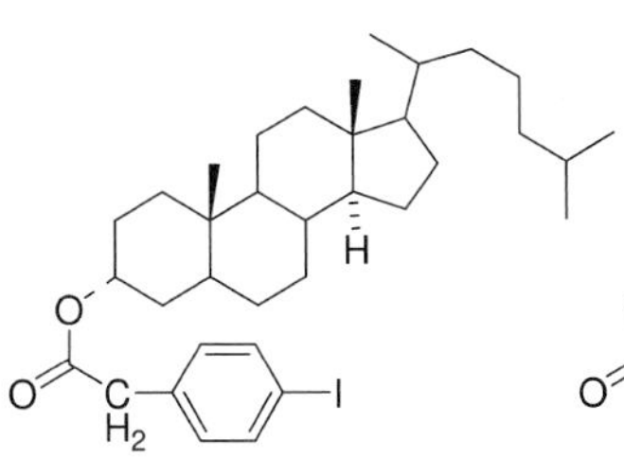

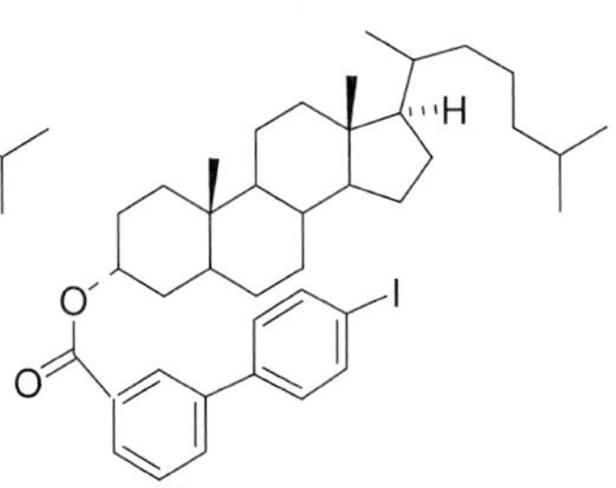

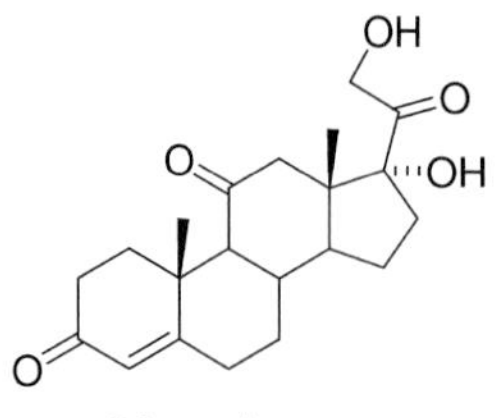

91, cortisone

moving by the time the rather tight transition state for hydrogen atom abstraction occurs. As with the enzymatic processes, the advantage is entropy.

We used the radical relay process, chlorinating C-9 and then generating the 9(11) double bond, in a synthesis of cortisone **91** [158]. This is a substitute for manufacturing processes in which C-9 or C-11 are hydroxylated by biological fermentation. Also, with templates that directed the chlorination to C-17 of 3α-cholestanol, such as that in **90**, we were able to remove the steroid sidechain [159–162]. Using an electrochemical oxidation process, we could direct chlorination by simple chloride *ion* with an iodophenyl template [163]. A general review of the processes with iodophenyl templates has been published [164].

The radical relay process also works with other template types. Thus, the thioether unit in **92** directed chlorination of C-14 by SO_2Cl_2 [165]. Also, the sulfur in the thioxanthone template of **93** directed the radical relay process to C-9 [166]. The thiophene sulfur in **94** was able to direct chlorination to C-9 in all three attached steroids [167]. In all these cases, an intermediate is formed with a chlorine atom bonded to sulfur.

A chlorine atom could also coordinate to the nitrogen of the pyridine template in compound **95**, directing chlorination to C-9 in a radical relay process [168]. Spectro-

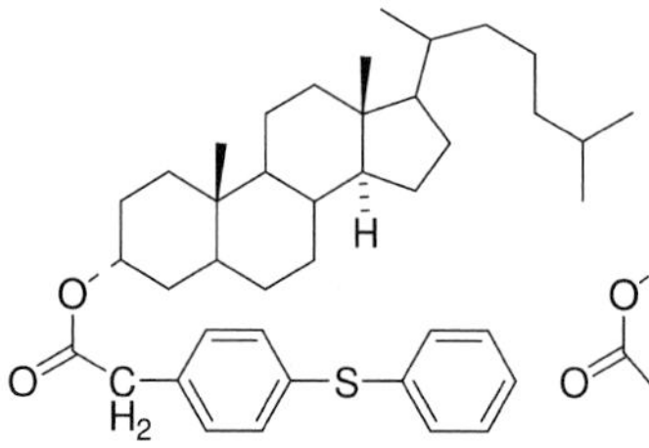

92

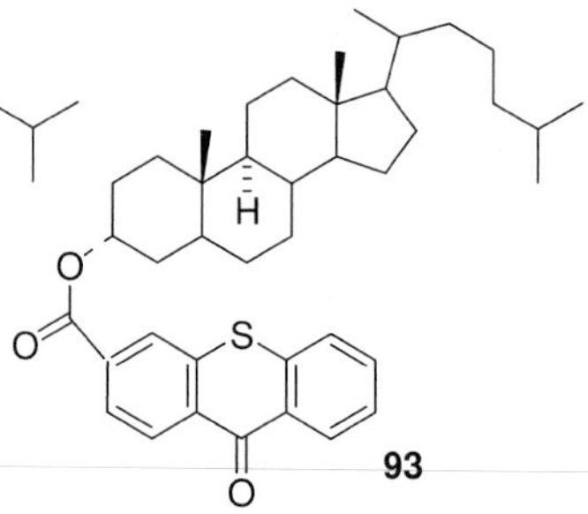

93

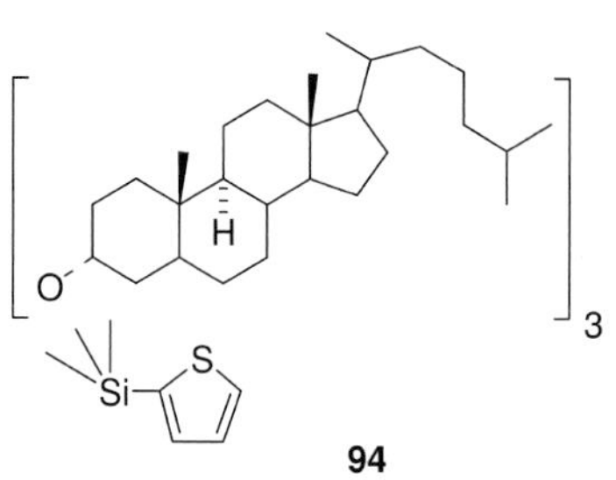

94

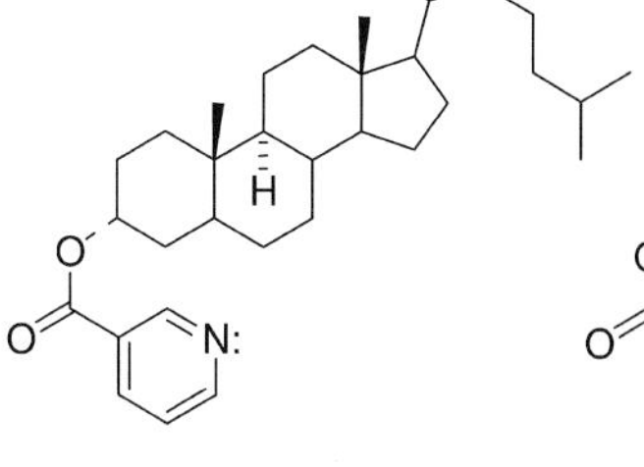

95 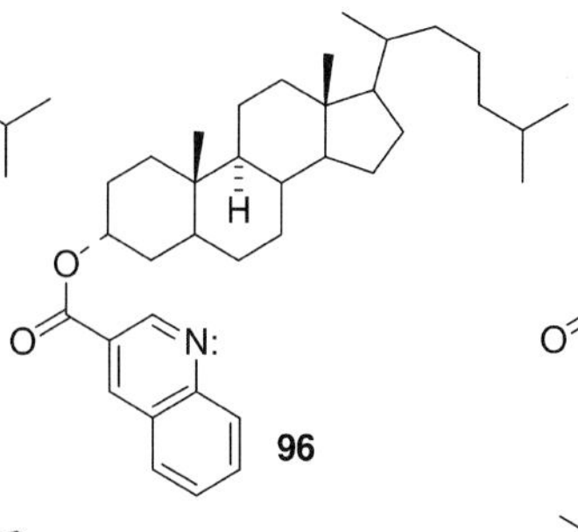   96 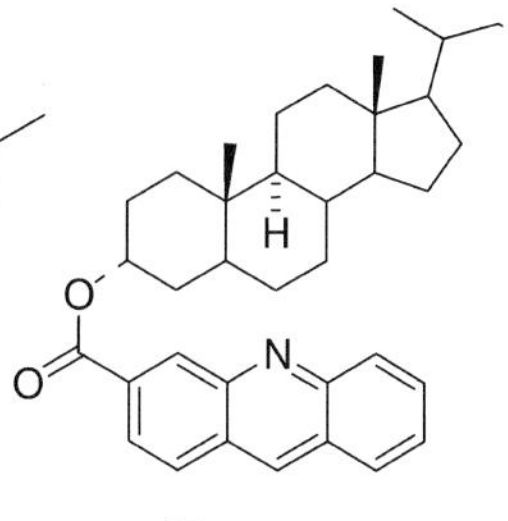   97

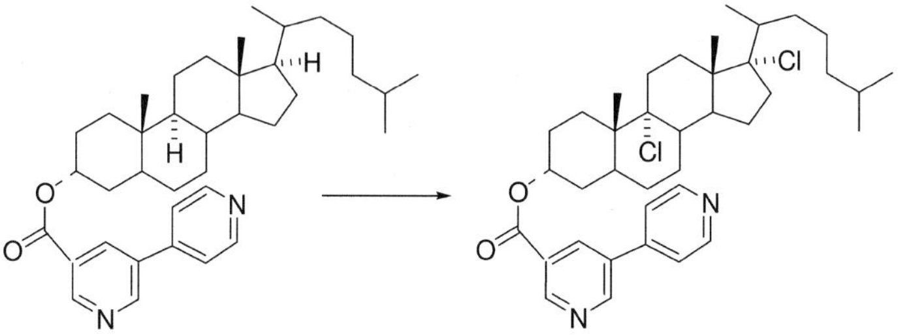

98   99

scopy and detailed quantum mechanical calculations indicated that the Cl–N bond is a localized three-electron coordination, strong enough to hold the chlorine in place but weak enough to permit it to attack tertiary C–H bonds [169]. Quinoline and acridine templates in **96** and **97**, respectively, could also use this nitrogen coordination of a chlorine atom to direct steroid chlorination [170]. In a particularly interesting example, both nitrogen positions in the template in **98** were used for chlorine coordination, producing the 9,17-dichloroproduct **99** in quantitative yield [171]. We have reviewed the use of heterocycles as radical relay templates [172].

The templates can be simply coordinated rather than attached. For example, complex **100** directed the radical relay chlorination to C-9, although the process was not as clean as with the attached templates [173]. We also used template-directed chlorinations to determine the conformations of flexible chains, just as we had previously with the benzophenone probes [174]. Also, by use of a set of tandem free radical chain reactions we could direct the formation of carbon–bromine and carbon–sulfur bonds, again with geometric control by the attached template [175].

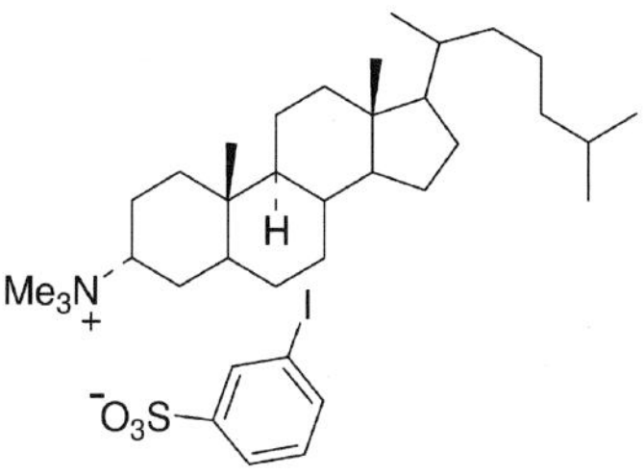

100

The template-directed steroid functionalizations are remarkably selective. After all, the C-9 and C-14 hydrogens are very close (1,3 diaxial on ring C). In a theoretical paper we discussed this selectivity, and concluded that an extra factor in the selectivity may be more than the ability of a chlorine atom to simply collide with one or the other of these hydrogens [176]. Instead, the trajectory of the collision may be important, requiring a straight C–H–Cl alignment. This would help select among such closely spaced hydrogens.

1.4.4
Nitrene Insertions

We briefly explored the use of nitrene insertions directed by geometric placement of an appropriate nitrene precursor, which was generated by photolysis of an azide. Although some nitrenes readily rearrange, we saw a class of phosphoryl nitrenes that performed intermolecular insertions rather than rearrangements [177]. However, when such a nitrene **102** was generated in cyclohexane solution by photolysis of **101** it preferentially attacked the solvent, forming **103**, rather than performing an intramolecular insertion reaction into the neighboring benzylic C–H bond [178]. Intramolecular product **105** was indeed seen when azide **101** was decomposed thermally in cyclohexane in the presence of a rhodium salt. Thus a free nitrene is so reactive that the normal preference for intramolecular reactions is not seen – it reacts on essentially every collision and is surrounded by solvent. Rhodium complexed nitrene **104**, a "nitrenoid", is less reactive and shows the entropy advantage of an intramolecular reaction with geometric control.

Such metal-complexed nitrenes were also generated by the reaction of (tosyliminoiodo)benzene (**106**) with Mn(III)- or Fe(II)-tetraphenylporphyrin, **107**, in a mimic of cytochrome P-450 but with a tosylimino group instead of an oxygen atom on the metals (**108**) [179]. It was able to functionalize cyclohexane solvent, by nitrogen insertion into a C–H bond to form **109**. Furthermore, the metalloporphyrins also catalyzed an intramolecular nitrogen insertion converting **110** into **111** [180].

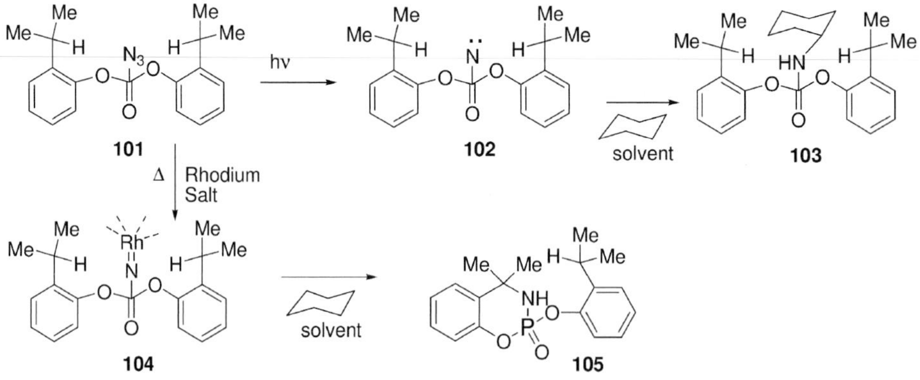

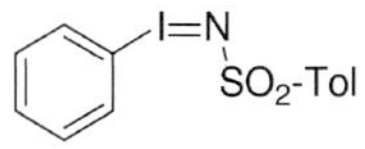

106
Ph-I=NTs

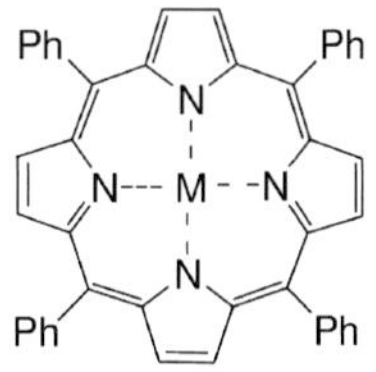

107, M = Mn(III)
or Fe(II)

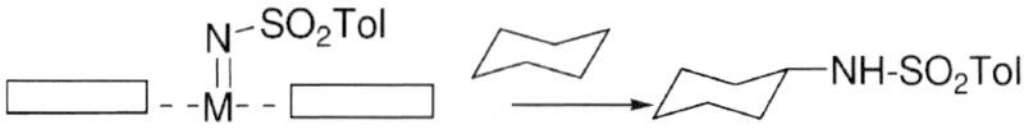

108
A metalloporphyrin
imine

109

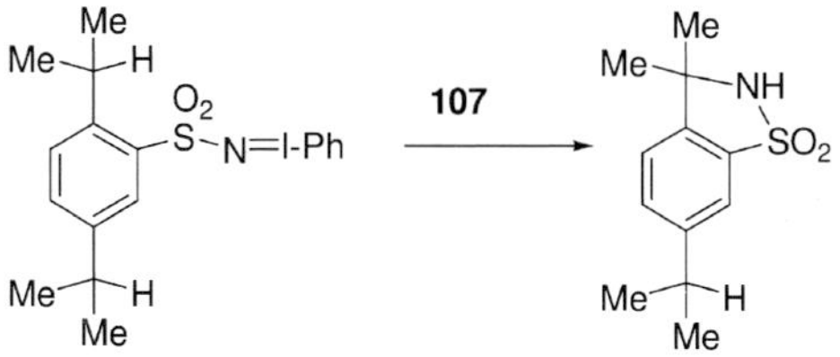

110

107

111

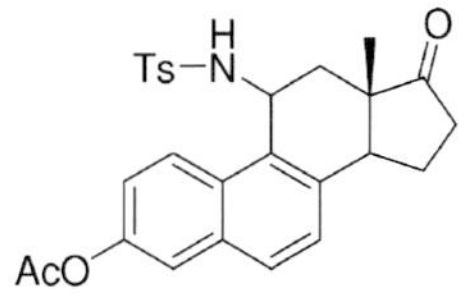

112

Cytochrome P-450 can perform both intramolecular and intermolecular nitrene insertions with **106** in water solution [181]. However, the intermediate metallo-nitrene (**108**) also hydrolyzed to some extent; so the enzyme performed a hydroxylation as well as an amidation. Although we have also performed an intermolecular amidation of a steroid, using such a metalloporphyrin reaction with **106** to form **112** [182], we have not yet extended it to the kinds of directed functionalizations described above with either benzophenones or chlorinations.

1.4.5
Binding by Cyclodextrin Dimers

To create artificial enzymes that could bind substrates in water solution with defined geometry, we examined dimers of cyclodextrins. As mentioned above, we used such dimers in mimics of hydrolytic enzymes [119, 120]. Now we wished to use them for mimics of cytochrome P-450.

Initially, we prepared β-cyclodextrin dimers **113–116** [183], which were examined for substrate binding with two p-t-butylphenyl groups. With ester **117** and cyclopropene **118** we saw binding constants as high as 10^8 M^{-1}. This is double the free energy of binding of simple p-t-butylphenyl groups into a single cyclodextrin – binding constants are generally about 10^4 M^{-1} – so the free energies of binding were additive with substrates that could fit well.

Even stronger dimeric binding, due to the entropy advantages of chelate binding, was seen with cyclodextrin dimer **119**, which was doubly linked, restricting its rotational freedom [184]. The flexible substrate **120** gave a binding constant to dimer **119** of 10^{10} M^{-1}, while with the rigid substrate **121** the binding constant was even larger. Interestingly, although the arguments for special effects in chelate binding normally involve *entropy* advantages, a study we did of several such chelate binding situations

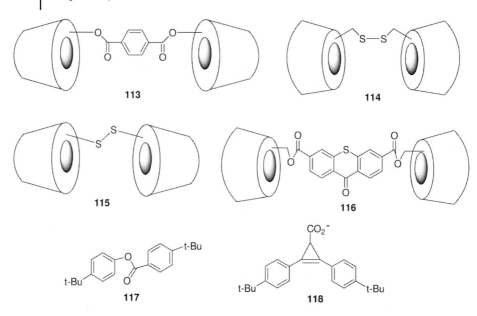

113   114

115   116

117   118

with cyclodextrin dimers and double-ended substrates showed that the special chelate effect was reflected in the *enthalpies* [185]. Entropy–enthalpy compensation is not an unusual phenomenon. We performed a number of such studies, including one in which we were able to quantitate the binding energy that could be ascribed to the hydrophobic effect [186].

We also investigated chelate binding by dimers of a synthetic hydrophobic macrocycle, in place of the cyclodextrins [187]. In the systems examined the chelate effect was weaker than that seen with the cyclodextrin dimers. We also studied the strong binding of cholesterol by some cyclodextrin dimers and a cyclodextrin polymer, and saw that the large sterol could occupy parts of two binding cavities [188].

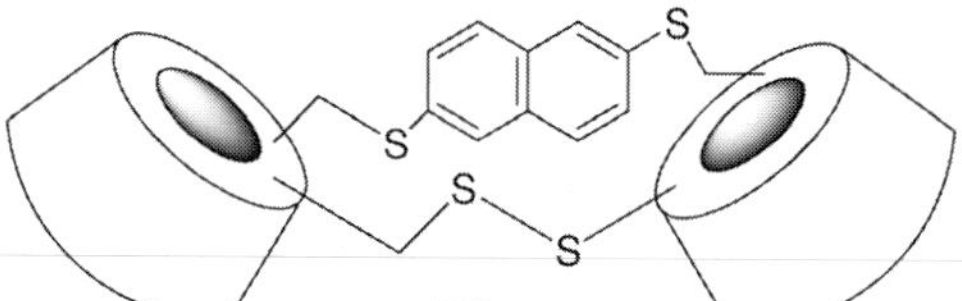

119

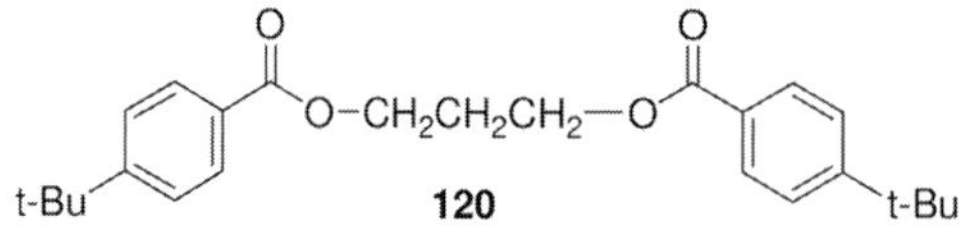

120 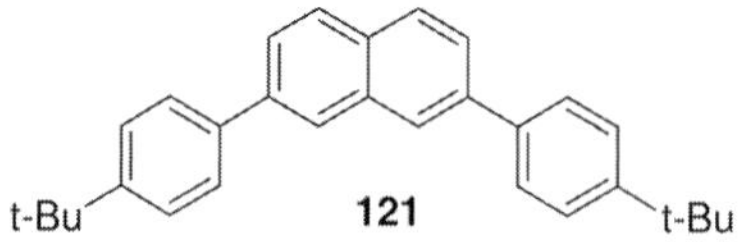   121

1.4.6
Hydroxylations by Artificial P-450 Enzymes

Metalloporphyrins can catalyze the hydroxylations of solvent species such as cyclohexane. From our studies with cyclodextrin dimers, we concluded that by attaching cyclodextrin rings to metalloporphyrins we should be able to bind substrates in water and achieve selective hydroxylations directed by the geometries of the complexes. This was successful.

Initially, instead of cyclodextrin complexing, we used metal coordination [189]. With Fe(III) porphyrin **122** carrying 8-hydroxyquinoline groups we epoxidized substrate **123** using Cu^{2+} as a bridge (cf. **124**). A metallosalen catalyst was not as effective. We then synthesized a series of Mn(III) porphyrins, **125–127**, as well as a Mn(III) metallosalen **128**, all carrying attached β-cyclodextrin groups [190]. We examined their ability to epoxidize the double bonds of substrates that could bind into two of the cyclodextrin

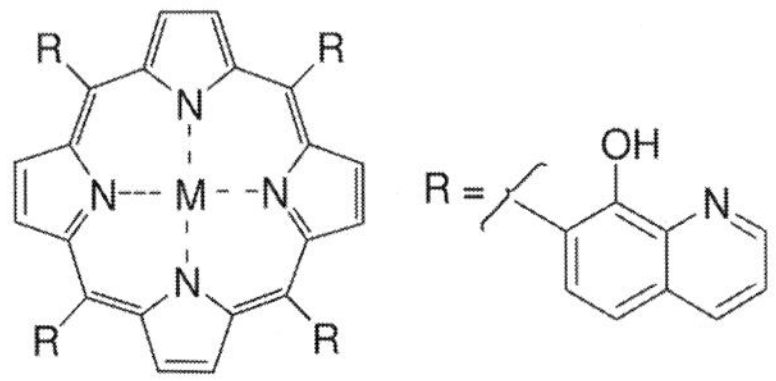

122, M = Fe(III)

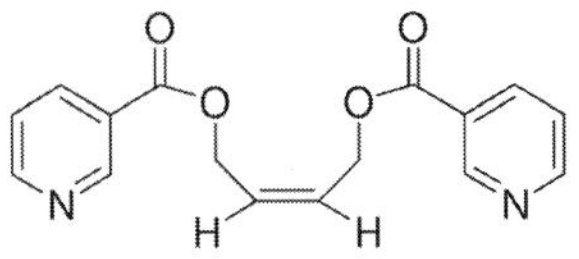

123

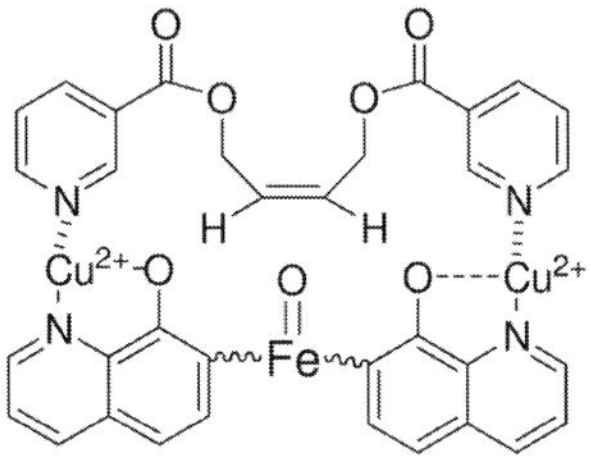

124

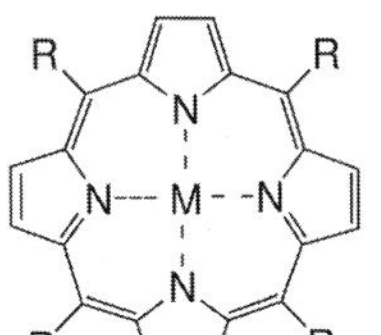

125, M = Mn(III) 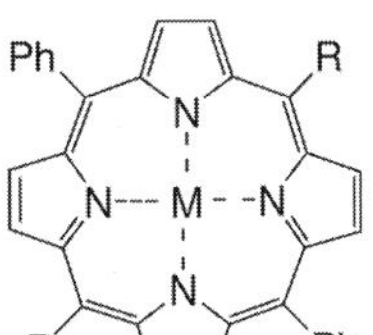   **126**, M = Mn(III) 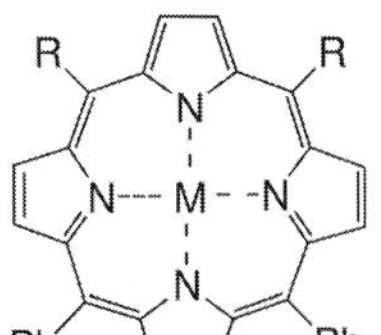   **127**, M = Mn(III)

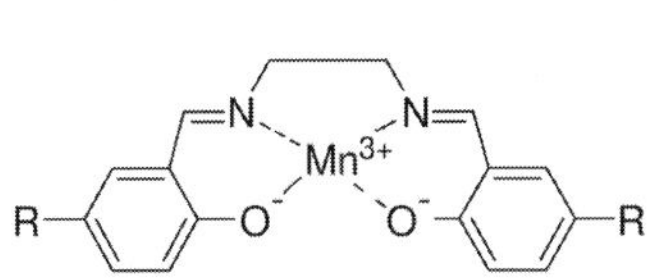

128

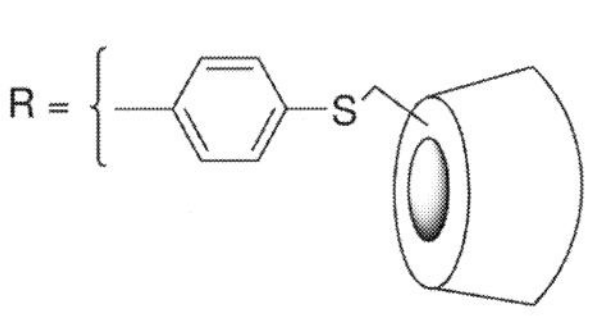

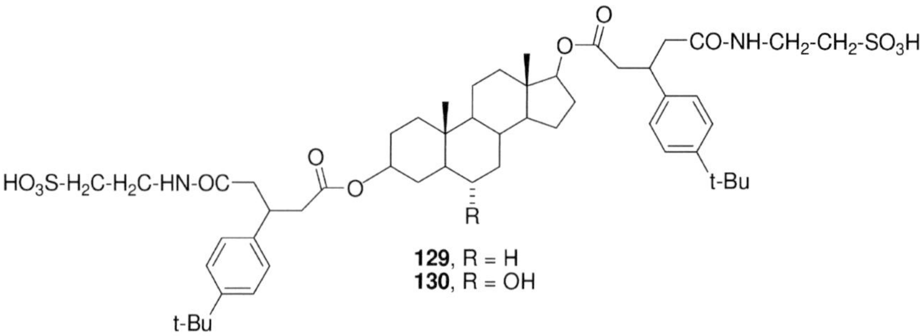

129, R = H
130, R = OH

groups, using iodosobenzene as the oxidant, and saw sensible variations in the relative rates of epoxidation with structure, indicating that the porphyrins that could bind the substrates across the face of the metalloporphyrin core were the most effective.

Building on these results, we examined the capability of manganese porphyrin **125** in hydroxylating steroid substrate **129**, again using iodosobenzene as oxidant [191, 192]. Indeed, the reaction was successful, and a single product was produced in which the C-6 equatorial hydrogen of substrate **129** was replaced with a hydroxyl group in product **130**. There was even some turnover catalysis, but with only 4 turnovers or so before the catalyst was oxidatively destroyed.

The turnover problem was solved with porphyrin derivative **131**, in which the fluorines on the phenyl groups greatly stabilized the catalyst against oxidative destruction [193]. Again with iodosobenzene as the reagent, catalyst **131** converted substrate **129** into its 6-hydroxy derivative **130**, but now with 187 turnovers. We have since produced catalysts with even higher turnovers for this process [194].

In our catalytic hydroxylations described above we added some pyridine to the solution to coordinate with one face of the manganese porphyrin, so as to direct an oxygen atom and the bound substrate to the same face. To avoid this, we synthesized catalyst

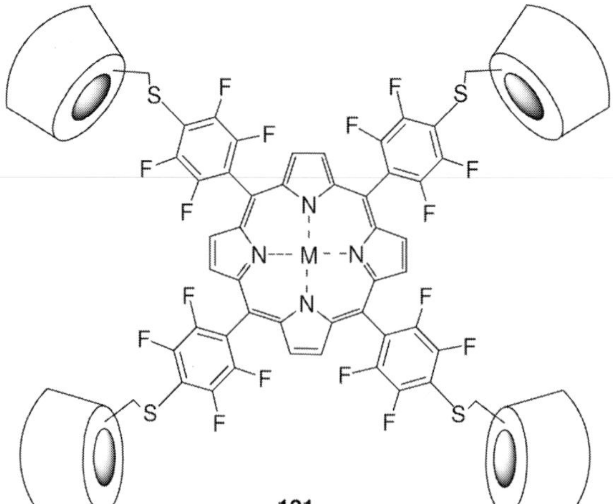

131

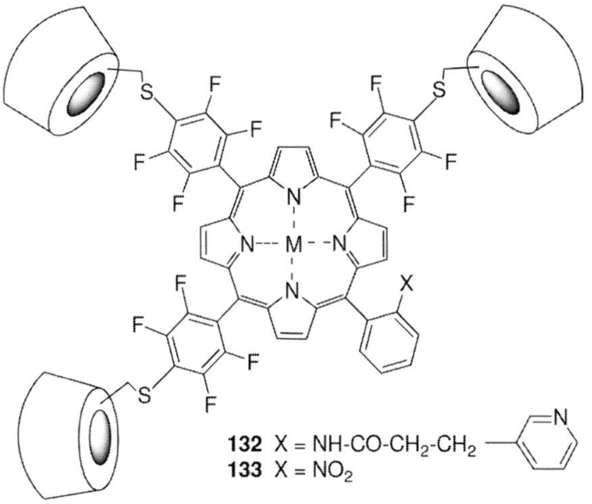

132 X = NH-CO-CH₂-CH₂ —
133 X = NO₂

132 with three cyclodextrins and a linked pyridine ring; no extra pyridine was needed with this catalyst [194]. As a bonus, this catalyst hydroxylated the C-6 position with 2000 turnovers. Catalyst **132** was synthesized through intermediate **133**, with a nitro-phenyl group, and this performed the C-6 hydroxylation with 3000 turnovers (although it still needed the added pyridine in solution). Such very high turnovers make these catalytic hydroxylations potentially practical.

While the selective hydroxylations at C-6 are certainly examples of geometrically directed functionalizations, imitating the enzyme cytochrome P-450, C-6 is not the most attractive target for such selectivity. We wanted to perform a selective multi-turnover hydroxylation of C-9 in the steroid. This would imitate the hydroxylations that take place in fermentations, and would let us convert the product into the 9(11) olefin, as we had done in our directed chlorinations described above.

Computer molecular modeling indicated that the hydroxylations at C-6 occurred because, with only two binding points for the substrate (into the two cyclodextrins on opposite sides of the porphyrin), the substrate could rotate so as to point its edge toward the Mn=O species that performs the hydroxylation. To solve this, we

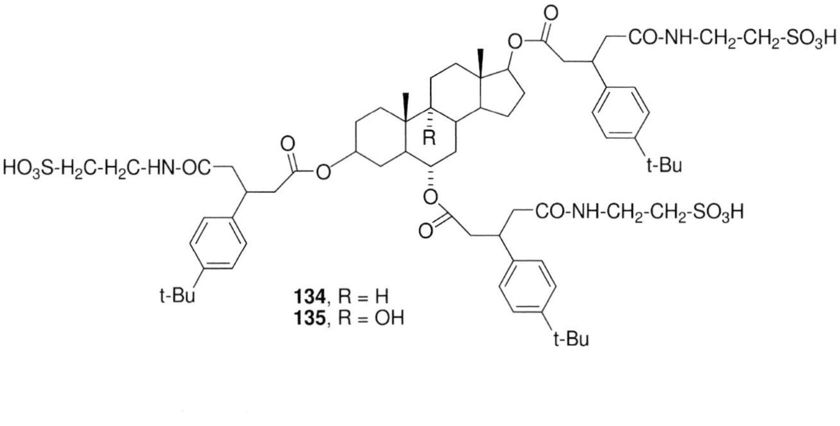

134, R = H
135, R = OH

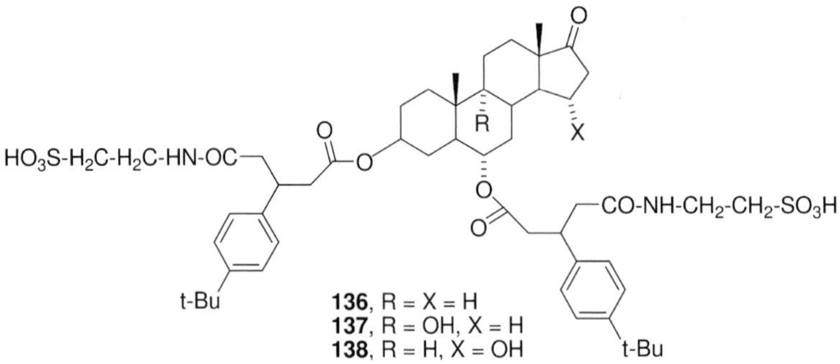

136, R = X = H
137, R = OH, X = H
138, R = H, X = OH

added an additional binding ester group to the substrate, in **134** [195]. This made *three* binding interactions with the catalyst **131**, and the hydroxylation product was now indeed the C-9 hydroxy compound **135**. Our computer models made it clear why this was successful [195]. Some of this work has been reviewed [196].

The third binding interaction introduced in the hydroxylation of **134** made the substrate present its face, not its edge, to the Mn=O of catalyst **131**, but it seemed likely that *double* binding of the substrate could also achieve this if the binding groups were attached to C-3 and C-6 rather than C-3 and C-17 as in **134**. Thus we synthesized a new substrate, **136**, and examined its hydroxylation by catalyst **131** [197]. We saw that indeed there was hydroxylation at C-9 to form **137**, but there was also some hydroxylation at C-15 to form **138**. With the lack of a binding interaction at C-17, apparently the substrate had slipped a little to the left, moving C-15 into position to be attacked.

To move the substrate back in place, by shortening the distance from a cyclodextrin to the Mn=O group, models suggested that the cyclodextrin be attached to the meta rather than para position of the phenyl rings in the catalyst. Thus we synthesized catalyst **139**, with such a meta attachment, and indeed it did cleanly convert substrate **136** into the C-9 hydroxy product **137**, with no hydroxylation at C-15 [197]. However, as there was no fluorine in the phenyl linkers, the catalyst was destroyed after only 2.5 turnovers.

This was solved by replacing the phenyl linkers with trifluoropyridine rings in catalyst **140** [197]. The compound was easily made, and it performed the C-9 hydroxylation of substrate **136** to form product **137** with 90 turnovers.

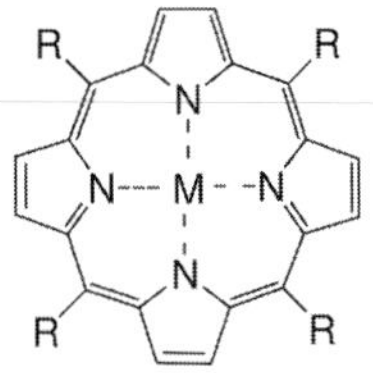

139, R =

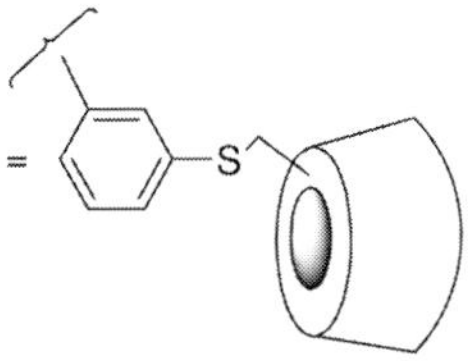

140, R =

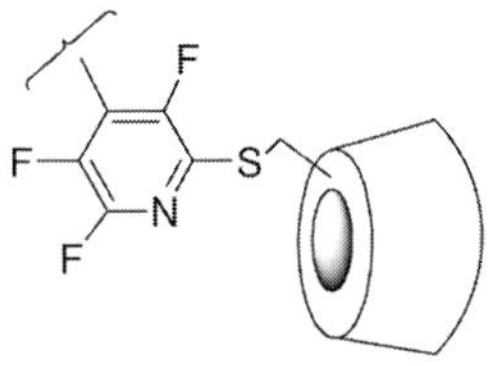

We have also synthesized a catalyst related to **131** in which the cyclodextrin rings were replaced with synthetic macrocyclic binding groups [196]. Also, we have examined catalysts related to **131** in which substrate binding involved metal ion coordination, not hydrophobic binding into cyclodextrins or macrocycles [198].

1.5
Future Prospects

As Philip Ball has pointed out, in biomimesis we take principles from Nature, not blueprints [199]. That is, we adopt the style of natural chemistry, but do not simply reproduce the same enzymes by which Nature achieves selectivity. His analogy is very apt: "A jumbo jet is not just a scaled-up pigeon." We learned the principle of wings from the birds, but not the details of how to use them and power the flight. Thus, in our work we have adopted the principles of reversible bonding and geometrically directed selectivity characteristic of natural enzymes, but used very different structures in our artificial enzymes. Our goal is to "liberate chemistry from the tyranny of functional groups" [194].

This new style of synthetic catalysis will of course not replace all normal synthetic methods. For many purposes, the standard methods and rules – e.g. aldehydes are more easily reduced than are ketones – will continue to dominate organic synthesis. However, when we require a synthetic transformation that is not accessible to normal procedures, as in the functionalization of unactivated carbons remote from functional groups, artificial enzymes can play a role. They must compete with natural enzymes, and with designed enzyme mutants, but for practical large-scale industrial synthesis there can be advantages with catalysts that are more rugged than proteins.

Our work in the development of artificial enzymes, described here and in Chapter 2, has established that catalysts can be made that will achieve excellent defined geometrically directed functionalizations and, furthermore, that the combination of binding groups with coenzyme analogs leads to powerful catalysts for some reactions. We expect the field of biomimetic chemistry to continue to grow, as we combine lessons from Nature with the ingenuity of chemists.

We will learn to produce mimics of enzyme clusters, imitating natural clusters such as gene transcription assemblies. We will learn to produce artificial enzymes that show induced fit, and allosteric control by analogs of hormones. Then we will move to mimics of cells themselves, with their components of many enzymes, to achieve chemical processes more complex than those done by a single enzyme. The biochemistry of life is impressive, but the role of chemistry is not just to admire it. As humans were impelled to invent ways to fly after observing birds, we will learn to create a new area of chemistry – biomimetic reaction chemistry – adding both to our understanding and to our practical abilities.

References

1 R. Breslow S. D. Dong, *Chem. Rev.* **1998**, *98*, 1997–2011.
2 R. Breslow **1971**, in *Bioinorganic Chemistry*, Advances in Chemistry Series, No. 100, American Chemical Society, Washington, 21–43,.
3 R. Breslow, Centenary Lecture, *Chem. Soc. Rev.* **1972**, *1*, 553–580.
4 R. Breslow **1978**, *Further Prospectives in Organic Chemistry*, Ciba Foundation Symposium 53, Elsevier Publishing Company, Amsterdam, 175–189.
5 R. Breslow, *Huaxue Tongbao* **1978**, *6*, 13.
6 R. Breslow, *J. Steroid Biochem.* **1979**, *11*, 19–26.
7 R. Breslow, *Israel J. Chem.* **1979**, *18*, 187–191.
8 R. Breslow, *Inf. Chim.* **1979**, *192–193*, 129–134.
9 R. Breslow, *Acc. Chem. Res.* **1980**, *13*, 170–177.
10 R. Breslow **1980**, in *Biomimetic Chemistry*, ed. D. Dolphin, C. McKenna, Y. Murakami, I. Tabushi, American Chemical Society, Washington, D.C., 1–15.
11 R. Breslow, *Science* **1982**, *218*, 532–537.
12 R. Breslow, *Chem. Br.* **1983**, *19*, 126–131.
13 R. Breslow **1983**, in *Biomimetic Chemistry*, ed. Z.-I. Yoshida, N. Ise, Kodansha Ltd. Publishing Co., Tokyo, 1–20.
14 R. Breslow, *2YC Distillate, Am. Chem. Soc.* **1983**, *1*, 6–7.
15 R. Breslow **1984**, in *Inclusion Compounds*, ed. J. L. Atwood, J. E. Davies, Academic Press, Inc., Orlando, Fla., Vol. 3, 473–508.
16 R. Breslow **1986**, in *Chemical Reactions in Organic and Inorganic Constrained Systems*, ed. R. Setton, D. Reidel, Dordrecht, l7–28.
17 R. Breslow **1986**, in *Advances in Enzymology and Related Areas of Molecular Biology*, ed. A. Meister, John Wiley & Sons, Inc., Vol. 58, 1–60.
18 R. Breslow, *Ann. N.Y Acad. Sci.* **1986**, *471*, 60–69.
19 R. Breslow **1986**, in *Proceedings of the XVIII Solvay Conference on Chemistry*, ed. G. van Binst, Springer-Verlag, Berlin, 185–197.
20 R. Breslow, *Cold Spring Harbor Symp. Quantitative Biol.* **1987**, *52*, 75–81.
21 R. Breslow, *Chem. Scr.* **1987**, *27*, 555–556.
22 R. Breslow, *New Sci.* **1988**, *119*, 44.
23 R. Breslow, *Proc. Robert A. Welch Found. Conf. Chem. Res.* **1988**, *31*, 73–89.
24 R. Breslow, *Chemtracts Org. Chem.* **1988**, *1*, 333–348.
25 R. Breslow, *UCLA Symp. Mol. Cell. Biol.* **1990**, *110*, 135–144.
26 R. Breslow, *Pure & Appl. Chem.* **1990**, *62*, 1859–1866.
27 R. Breslow, *Ciba Foundation Symp.* **1991**, *158*, 115–127.
28 R. Breslow, *Acc. Chem. Res.* **1991**, *24*, 317–324.
29 R. Breslow **1991**, in *Comprehensive Organic Synthesis*, ed. B. M. Trost, Pergamon Press, Oxford, Vol. 7, 39–52.
30 R. Breslow, *Supramol. Chem.* **1992**, 411–428.
31 R. Breslow, *Israel J. Chem.* **1992**, *32*, 23–30.
32 R. Breslow **1992**, *Minutes of the Sixth International Symposium on Cyclodextrins*, Editions de Sante, Paris, 625–630.
33 R. Breslow, *Proc. Natl. Acad. Sci. USA* **1993**, *90*, 1139.
34 R. Breslow, *Supramol. Chem.* **1993**, *1*, 111–118.
35 R. Breslow, *Pure & Appl. Chem.* **1990**, *66*, 1573–1582.
36 R. Breslow, *J. Mol. Cat.* **1994**, *91*, 161–174.
37 R. Breslow, *Recl. Trav. Chim. Pays-Bas* **1994**, *113*, 493–498.
38 R. Breslow, *Acc. Chem. Res.* **1995**, *28*, 146–153.
39 R. Breslow **1995**, in *Organic Reactivity: Physical and Biological Aspects*, ed. B. T. Golding, R. J. Griffin, and H. Maskill, The Royal Society of Chemistry, Cambridge, Special Publication No. 148, 3–24.
40 R. Breslow, *Supramol. Chem.* **1995**, *66*, 41–47.
41 R. Breslow, *NATO ASI Ser., Ser. E* **1996**, *320*, 113–135.
42 R. Breslow, *Proc. Robert A Welch Foundation 40th Conference on Chem. Res.* **1997**, *40*, 1–11.
43 R. Breslow, *Chem. & Biol.* **1998**, *5*, R27–28.
44 R. Breslow, *J. Chem. Ed.* **1998**, *75*, 705–718.
45 R. Breslow, *Pure & Appl. Chem.* **1998**, *70*, 267–270.
46 R. Breslow **1999**, in *Templated Organic Synthesis*, ed. F. Diederich and P. Stang, Wiley-VCH, Weinheim, Ch. 6.
47 R. Breslow, *Proc. Am. Phil. Soc.*, **2000**, *144*, 267–274.
48 R. Breslow, *Chem. Record*, **2001**, *1*, 3–11.
49 R. Breslow, *Biol. J. Armenia* **2001**, *53*, 117–120.
50 R. Breslow, *Chemtracts Org. Chem.*, **2002**, *15*, 59–68.
51 R. Breslow, *Polym. Preprints*, **2003**, *44*, 480.
52 R. Breslow, *Chem. Ind., BIF Rev.*, **1956**, R.28–29.
53 R. Breslow, *J. Am. Chem. Soc.* **1957**, *79*, 1762.
54 R. Breslow, *Chem. & Ind.* **1957**, 893–894.
55 R. Breslow, *J. Am. Chem. Soc.* **1958**, *80*, 3719–3726.

56 R. Breslow, E. McNelis, *J. Am. Chem. Soc.* **1959**, *81*, 3080–3082.

57 R. Breslow, E. McNelis, *J. Am. Chem. Soc.* **1960**, *82*, 2394.

58 R. Breslow, *CIBA Foundation Study Group II*, J.A. Churchill Ltd., **1961**, p. 65.

59 R. Breslow, *Ann. N.Y. Acad. Sci.* **1962**, *98*, 445–452.

60 R. Breslow, C. Schmuck, *Tetrahedron Lett.* **1996**, *37*, 8241–8242.

61 F. Lopez-Calahorra, R. Rubires, *Tetrahedron* **1995**, *51*, 9713.

62 F. Lopez-Calahorra, E. Castro, A. Ochoa, J. Marti, *Tetrahedron Lett.* **1996**, *37*, 5019.

63 D. Hilvert, R. Breslow, *Bioorg. Chem.* **1984**, *12*, 206–220.

64 R. Breslow, E. Kool, *Tetrahedron Lett.* **1988**, *29*, 1635–1638.

65 R. Breslow, K. Groves, M. U. Mayer, *Pure & Appl. Chem.* **1998**, *70*, 1933–1938.

66 R. Breslow, K. Groves, M. U. Mayer, *J. Am. Chem. Soc.* **2002**, *124*, 3622–3635.

67 R. Breslow, M. Hammond, M. Lauer, *J. Am. Chem. Soc.* **1980**, *102*, 421–422.

68 R. Breslow, A. W. Czarnik, *J. Am. Chem. Soc.* **1983**, *105*, 1390–1391.

69 A. W. Czarnik, R. Breslow, *Carbohydr. Res.* **1984**, *128*, 133–139.

70 R. Breslow, J. Chmielewski, D. Foley, B. Johnson, N. Kumabe, M. Varney, R. Mehra, *Tetrahedron* **1988**, *44*, 5515–5524.

71 R. Breslow, J. W. Canary, M. Varney, S. T. Waddell, D. Yang, *J. Am. Chem. Soc.* **1990**, *112*, 5212–5219.

72 J. Winkler, E. Coutouli-Argyropoulou, R. Leppkes, R. Breslow, *J. Am. Chem. Soc.* **1983**, *105*, 7198–7199.

73 S. C. Zimmerman, A. W. Czarnik, R. Breslow, *J. Am. Chem. Soc.* **1983**, *105*, 1694–1695.

74 S. C. Zimmerman, R. Breslow, *J. Am. Chem. Soc.* **1984**, *106*, 1490–1491.

75 R. Breslow, A. W. Czarnik, M. Lauer, R. Leppkes, J. Winkler, S. Zimmerman, *J. Am. Chem. Soc.* **1986**, *108*, 1969–1979.

76 W. Weiner, J. Winkler, S. C. Zimmerman, A. W. Czarnik, R. Breslow, *J. Am. Chem. Soc.* **1985**, *107*, 4093–4094.

77 J. Chmielewski, R. Breslow, *Heterocycles* **1987**, *25*, 533.

78 J. T. Koh, L. Delaude, R. Breslow, *J. Am. Chem. Soc.* **1994**, *116*, 11234–11240.

79 L. Liu, R. Breslow, *Tetrahedron Lett.* **2001**, *42*, 2775–2777.

80 L. Liu, M. Rozenman, R. Breslow, *Bioorg. Med. Chem.* **2002**, *10*, 3973–3979.

81 R. Breslow, S. Chung, *Tetrahedron Lett.* **1989**, *30*, 4345–4346.

82 R. Breslow, *J. Chem. Ed.* **1990**, *67*, 228–229.

83 D. W. Griffiths, M. L. Bender, *Adv. Catal.* **1973**, *23*, 209.

84 J. Emert, R. Breslow, *J. Am. Chem. Soc.* **1975**, *97*, 670–672.

85 M. F. Czarniecki. R. Breslow, *J. Am. Chem. Soc.* **1978**, *100*, 7771–7772.

86 R. Breslow, M. F. Czarniecki, J. Emert, H. Hamaguchi, *J. Am. Chem. Soc.* **1980**, *102*, 762–770.

87 G. Trainor, R. Breslow, *J. Am. Chem. Soc.* **1981**, *103*, 154–158.

88 R. Breslow, G. Trainor, A. Ueno, *J. Am. Chem. Soc.* **1983**, *105*, 2739–2744.

89 B. Siegel, R. Breslow, *J. Am. Chem. Soc.* **1975**, *97*, 6869–6870.

90 W.J. le Noble, S. Srivastava, R. Breslow, G. Trainor, *J. Am. Chem. Soc.* **1983**, *105*, 2745–2748.

91 H.-J. Thiem, M. Brandl, R. Breslow, *J. Am. Chem. Soc.* **1988**, *110*, 8612–8616.

92 R. Breslow, S. Chung, *Tetrahedron Lett.* **1990**, *31*, 631–634.

93 B. Siegel, A. Pinter, R. Breslow, *J. Am. Chem. Soc.* **1977**, *99*, 2309–2312.

94 R. Breslow, *Chem. & Eng. News*, September 8, **2003**, 86.

95 R. Breslow, D. Wernick, *J. Am. Chem. Soc.* **1976**, *98*, 259–261.

96 R. Breslow, D. L. Wernick, *Proc. Natl. Acad. Sci. USA*, **1977**, *74*, 1303–1307.

97 R. Breslow, J. Chin, D. Hilvert, *Inorg. Chim. Acta* **1983**, *79*, 61.

98 R. Breslow, J. Chin, D. Hilvert, G. Trainor, *Proc. Nat. Acad. Sci. USA* **1983**, *80*, 4585–4589.

99 R. Breslow, A. Schepartz, *Chem. Lett.* **1987**, 1–4.

100 D. W. Christianson, W. N. Lipscomb, *Acc. Chem. Res.* **1989**, *22*, 62–69.

101 R. Breslow, D. Chipman, *J. Am. Chem. Soc.* **1965**, *87*, 4195–4196.

102 R. Breslow, L. E. Overman, *J. Am. Chem. Soc.* **1970**, *92*, 1075–1077.

103 R. Breslow, I. Katz, *J. Am. Chem. Soc.* **1968**, *90*, 7376.

104 R. Breslow, R. Fairweather, J. Keana, *J. Am. Chem. Soc.* **1967**, *89*, 2135–2138.

105 R. Breslow, M. Schmir, *J. Am. Chem. Soc.* **1971**, *93*, 4960–4961.

106 R. Breslow, C. McAllister, *J. Am. Chem. Soc.* **1971**, *93*, 7096–7097.

107 R. Breslow, D. E. McClure, R. S. Brown, J. Eisenach, *J. Am. Chem. Soc.* **1975**, *97*, 194–195.

108 J. Chin, R. Breslow, *Tetrahedron Lett.* **1982**, *23*, 4221–4224.

109 R. Breslow, D. E. McClure, *J. Am. Chem. Soc.* **1976**, *98*, 258–259.

110 A. Schepartz, R. Breslow, *J. Am. Chem. Soc.* **1987**, *109*, 1814–1826.

111 J. Light, R. Breslow, *Bioorg. Chem.* **1990**, *18*, 63–77.

112 R. Breslow, B. Zhang, *J. Am. Chem. Soc.* **1992**, *114*, 5882–5883.

113 B. Zhang, R. Breslow, *J. Am. Chem. Soc.* **1997**, *119*, 1676–1681.

114 R. Breslow, S. Belvedere, L. Gershell, D. Leung, *Pure Appl. Chem.* **2000**, *72*, 333–342.

115 R. Breslow, S. Belvedere, L. Gershell, *Helv. Chim. Acta* **2000**, *83*, 1685–1692.

116 J. Yan, R. Breslow, *Tetrahedron Lett.* **2000**, *41*, 2059–2062.

117 R. Breslow, N. Nesnas, *Tetrahedron Lett.* **1999**, *40*, 3335–3338.

118 S. Gellman, R. Petter, R. Breslow, *J. Am. Chem. Soc.*, **1986**, *108*, 2388–2394.

119 R. Breslow, S. Singh, *Bioorg. Chem.* **1988**, *16*, 408–417.

120 R. Breslow, B. Zhang, *J. Am. Chem. Soc.* **1994**, *116*, 7893–7894.

121 R. Breslow, D. Berger, D.-L. Huang, *J. Am. Chem. Soc.* **1990**, *112*, 3686–3687.

122 S. D. Dong, R. Breslow, *Tetrahedron Lett.* **1998**, *39*, 9343–9346.

123 R. Breslow, D.-L. Huang, *Proc. Natl. Acad. Sci. USA* **1991**, *88*, 4080–4083.

124 R. Breslow, J. Doherty, G. Guillot, C. Lipsey, *J. Am. Chem. Soc.* **1978**, *100*, 3227–3229.

125 R. Breslow, P. Bovy, C. Lipsey Hersh, *J. Am. Chem. Soc.* **1980**, *102*, 2115–2117.

126 R, Corcoran, M. Labelle, A. W. Czarnik, R. Breslow, *Anal. Biochem.* **1985**, *144*, 563–568.

127 R. Breslow, M. Labelle, *J. Am. Chem. Soc.* **1986**, *108*, 2655–2659.

128 E. Anslyn, R. Breslow, *J. Am. Chem. Soc.* **1989**, *111*, 4473–4482.

129 E. Anslyn, R. Breslow, *J. Am. Chem. Soc.* **1989**, *111*, 5972–5973.

130 E. Anslyn, R. Breslow, *J. Am. Chem. Soc.* **1989**, *111*, 8931–8932.

131 R. Breslow, E. Anslyn, D.-L. Huang, *Tetrahedron* **1991**, *47*, 2365–2376.

132 R. Breslow, R. Xu, *Proc. Natl. Acad. Sci. USA* **1993**, *90*, 1201–1207.

133 R. Breslow, *Proc. Natl. Acad. Sci. USA* **1993**, *90*, 1208–1211.

134 R. Breslow, R. Xu, *J. Am. Chem. Soc.* **1993**, *115*, 10705–10713.

135 R. Breslow, S. D. Dong, Y. Webb, R. Xu, *J. Am. Chem. Soc.* **1996**, *118*, 6588–6600.

136 R. Breslow, C. Schmuck, *J. Am. Chem. Soc.* **1996**, *118*, 6601–6605.

137 R. Breslow, W. H. Chapman, Jr., *Proc. Natl. Acad. Sci. USA*, **1996**, *93*, 10018–10021.

138 R. Breslow, A. Graff, *J. Am. Chem. Soc.* **1993**, *115*, 10988–10989.

139 J. M. Desper, R. Breslow, *J. Am. Chem. Soc.* **1994**, *116*, 12081–12082.

140 R. Breslow, J. Desper, Y. Huang, *Tetrahedron Lett.* **1996**, *37*, 2541–2544.

141 D.-Q. Yuan, S. D. Dong, R. Breslow, *Tetrahedron Lett.* **1998**, *39*, 7673–7676.

142 R. Breslow, P. Campbell, *J. Am. Chem. Soc.* **1969**, *91*, 3085.

143 R. Breslow, P. Campbell, *Bioorg. Chem.* **1971**, *1*, 140–156.

144 R. Breslow, H. Kohn, B. Siegel, *Tetrahedron Lett.* **1976**, 1645–1646.

145 R. Breslow, M. A. Winnik, *J. Am. Chem. Soc.* **1969**, *91*, 3083–3084.

146 R. Breslow, J. Rothbard, F. Herman, M. L. Rodriguez, *J. Am. Chem. Soc.* **1978**, *100*, 1213–1218.

147 R. Breslow, S. Kitabatake, J. Rothbard, *J. Am. Chem. Soc.* **1978**, *100*, 8156–8160.

148 M. F. Czarniecki, R. Breslow, *J. Am. Chem. Soc.* **1979**, *101*, 3675–3676.

149 R. Breslow, S. W. Baldwin, *J. Am. Chem. Soc.* **1970**, *92*, 732–734.

150 R. Breslow, S. Baldwin, T. Flechtner, P. Kalicky, S. Liu, W. Washburn, *J. Am. Chem. Soc.* **1973**, *95*, 3251–3262.

151 R. Breslow, P. C. Scholl, *J. Am. Chem. Soc.* **1971**, *93*, 2331–2333.

152 R. Breslow, U. Maitra, D. Heyer, *Tetrahedron Lett.* **1984**, *25*, 1123–1126.

153 R. L. Wife, D. Prezant, R. Breslow, *Tetrahedron Lett.* **1976**, 517–520.

154 R. Breslow, R. Rajagopalan, J. Schwarz, *J. Am. Chem. Soc.* **1981**, *103*, 2905–2907.

155 R. Breslow, J. A. Dale, P. Kalicky, S. Y. Liu, W. N. Washburn, *J. Am. Chem. Soc.* **1972**, *94*, 3276–3278.

156 R. Breslow, R. Corcoran, J. A. Dale, S. Liu, P. Kalicky, *J. Am. Chem. Soc.* **1974**, *96*, 1973–1974.

157 R. Breslow, R. J. Corcoran, B. B. Snider, *J. Am. Chem. Soc.* **1974**, *96*, 6791–6792.

158 R. Breslow, B. B. Snider, R. J. Corcoran, *J. Am. Chem. Soc.* **1974**, *96*, 6792–6794.

159 B. B. Snider, R. J. Corcoran, R. Breslow, *J. Am. Chem. Soc.* **1975**, *97*, 6580–6581.

160 R. Breslow, U. Maitra, *Tetrahedron Lett.* **1984**, *25*, 5843–5846.

161 U. Maitra, R. Breslow, *Tetrahedron Lett.* **1986**, *27*, 3087–3090.

162 R. Breslow, T. Link, *Tetrahedron Lett.* **1992**, *33*, 4145–4148.

163 R. Breslow, R. Goodin, *Tetrahedron Lett.* **1976**, *31*, 2675–2676.

164 R. Breslow, R. J. Corcoran, B. B. Snider, R. J. Doll, P. L. Khanna, R. Kaleya, *J. Am. Chem. Soc.* **1977**, 99, 905–915.

165 R. Breslow, R. L. Wife, D. Prezant, *Tetrahedron Lett.* **1976**, 1925–1926.

166 R. Breslow, T. Guo, *Tetrahedron Lett.* **1987**, *28*, 3187–3188.

167 R. Breslow, D. Heyer, *J. Am. Chem. Soc.* **1982**, *104*, 2045–2046.

168 R. Breslow, M. Brandl, J. Hunger, A. D. Adams, *J. Am. Chem. Soc.* **1987**, *109*, 3799–3801.

169 R. Breslow, M. Brandl, J. Hunger, N. J. Turro, K. Cassidy, K. Krogh-Jespersen, J. D. Westbrook, *J. Am. Chem. Soc.* **1987**, *109*, 7204–7026.

170 R. Breslow, D. Wiedenfeld, *Tetrahedron Lett.* **1993**, *34*, 1107–1110.

171 R. Batra, R. Breslow, *Tetrahedron Lett.* **1989**, *30*, 535–538.

172 R. Breslow, A. Adams, M. Brandl, T. Guo, J. Hunger, *Lectures Heterocycl. Chem.* **1987**, *IX*, 43–49.

173 R. Breslow, D. Heyer, *Tetrahedron Lett.* **1983**, *24*, 5039–5042.

174 R. Batra, R. Breslow, *Heterocycles* **1989**, *28*, 23–28.

175 D. Wiedenfeld, R. Breslow, *J. Am. Chem. Soc.* **1991**, *113*, 8977–8978.

176 P. White, R. Breslow, *J. Am. Chem. Soc.* **1990**, *112*, 6842–6847.

177 R. Breslow, A. Feiring, F. Herman, *J. Am. Chem. Soc.* **1974**, *96*, 5937–5939.

178 R. Breslow, F. Herman, A. W. Schwabacher, *J. Am. Chem. Soc.* **1984**, *106*, 5359.

179 R. Breslow, S. H. Gellman, *J. Chem. Soc., Chem. Comm.* **1982**, 1400–1401.

180 R. Breslow, S. H. Gellman, *J. Am. Chem. Soc.* **1983**, *105*, 6728–6729.

181 E. W. Svastits, J. H. Dawson, R. Breslow, S. Gellman, *J. Am. Chem. Soc.* **1985**, *107*, 6427–6428.

182 J. Yang, R. Weinberg, R. Breslow, *Chem. Commun.* **2000**, 531–532.

183 R. Breslow, N. Greenspoon, T. Guo, R. Zarzycki, *J. Am. Chem. Soc.* **1989**, *111*, 8296–8297.

184 R. Breslow, S. Chung, *J. Am. Chem. Soc.* **1990**, *112*, 9659–9660.

185 B. Zhang, R. Breslow, *J. Am. Chem. Soc.* **1993**, *115*, 9353–9354.

186 R. Breslow, S. Halfon, B. Zhang, *Tetrahedron*, **1995**, *51*, 377–388.

187 R. Breslow, P. J. Duggan, D. Wiedenfeld, S. T. Waddell, *Tetrahedron Lett.* **1995**, *36*, 2707–2710.

188 R. Breslow, B. Zhang, *J. Am. Chem. Soc.* **1996**, *118*, 8495–8496.

189 R. Breslow, A. B. Brown, R. D. McCullough, P. W. White, *J. Am. Chem. Soc.* **1989**, *111*, 4517–4518.

190 R. Breslow, X. Zhang, R. Xu, M. Maletic, R. Merger, *J. Am. Chem. Soc.* **1996**, *118*, 11678–11679.

191 R. Breslow, X. Zhang, Y. Huang, *J. Am. Chem. Soc.* **1997**, *119*, 4535–4536.

192 R. Breslow, Y. Huang, X. Zhang, J. Yang, *Proc. Natl. Acad. Sci. USA* **1997**, *94*, 11156–11158.

193 R. Breslow, B. Gabriele, J. Yang, *Tetrahedron Lett.* **1998**, *39*, 2887–2890.

194 R. Breslow, J. Yang, J. Yan, *Tetrahedron* **2002**, *58*, 653–659.

195 J. Yang, R. Breslow, *Angew. Chem. Int. Ed.* **2000**, *39*, 2692–2694.

196 J. Yang, B. Gabriele, S. Belvedere, Y. Huang, R. Breslow, *J. Org. Chem.* **2002**, *67*, 5057–5067.

197 R. Breslow, J. Yan, S. Belvedere, *Tetrahedron Lett.* **2002**, *43*, 363–365.

198 S. Belvedere, R. Breslow, *Bioorg. Chem.* **2001**, *29*, 321–331.

199 P. Ball, *Stories of the Invisible*, Oxford University Press, Oxford, **2001**.

2
Vitamin B6 Enzyme Models

Lei Liu and Ronald Breslow

2.1
Introduction

Pyridoxal 5′-phosphate (PLP) and pyridoxamine 5′-phosphate (PMP) (Figure 2.1) are the active forms of vitamin B6 that function as cofactors for amino acid metabolizing enzymes [1]. These enzymes catalyze various reactions: (i) transamination between an α-amino acid and an α-keto acid; (ii) racemization of a chiral α-amino acid; (iii) elimination of an electronegative group at the β- or γ-position of an α-amino acid to give the corresponding α-keto acid and ammonia; (iv) replacement of an electronegative group at the β- or γ-position of an α-amino acid or keto acid with another substituent; (v) decarboxylation of an α-amino acid to give an amine; (vi) aldolase-type reaction accompanied with a C–C bond cleavage at the α,β-position of a β-hydroxy-α-amino acid.

In vitamin B6-dependent enzymes, PLP is invariably bound to an active site lysine through a Schiff base linkage, the internal aldimine. Its 5′-phosphate group is firmly attached to the protein matrix through up to nine hydrogen bonds, and often by charge interactions [2]. An external aldimine Schiff base, derived by transimination of the internal aldimine with a substrate amino acid, is a common intermediate for all vitamin B6-dependent enzymatic reactions. Due to the electron-sink effect of the conjugated pyridine segment, one of the three bonds around the α-carbon atom of the amino acid moiety in the intermediate undergoes cleavage (Scheme 2.1) [3]. Through

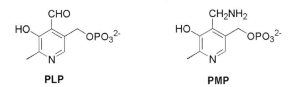

Figure 2.1 Pyridoxal 5′-phosphate and pyridoxamine 5′phosphate.

* Note: This chapter reviews both the work done in the Breslow laboratory and that done elsewhere. When "we" is used in the chapter it refers to work in the Breslow laboratory generally, not just that performed by Lei Liu. The full list of the authors involved in each case appears in the references.

Artificial Enzymes. Edited by R. Breslow
Copyright © 2005 WILEY-VCH Verlag GmbH & Co. KGaA, Weinheim
ISBN: 3-527-31165-3

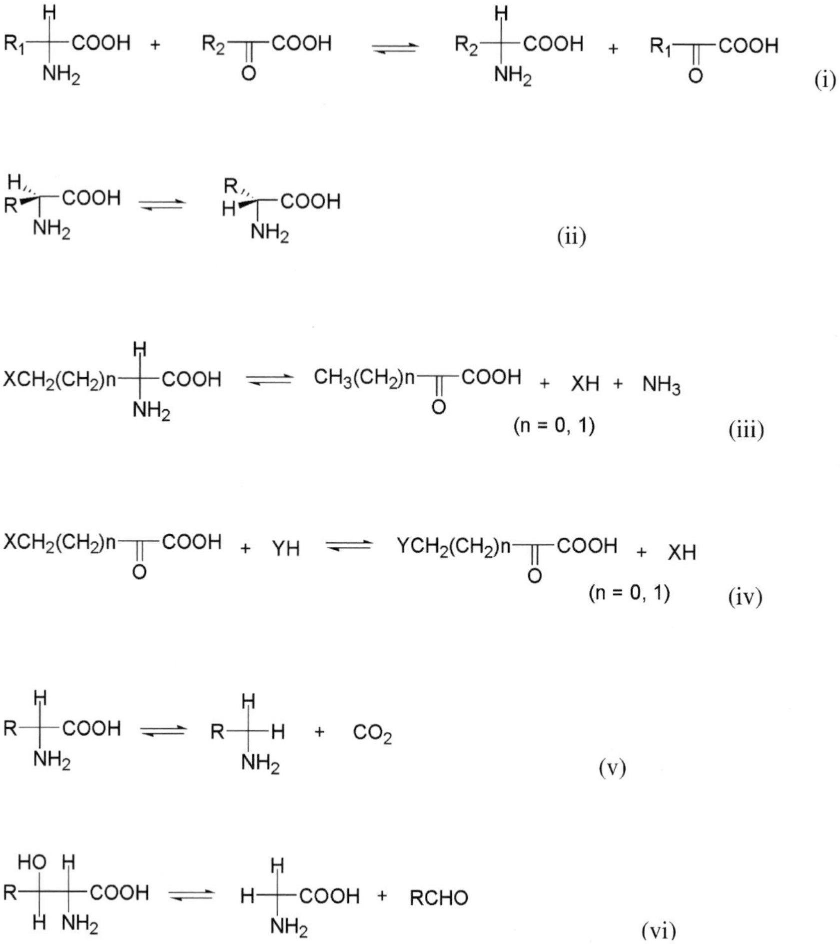

$$R_1 \underset{NH_2}{\overset{H}{\underset{|}{\overset{|}{C}}}} COOH \; + \; R_2 \underset{O}{\overset{}{\underset{\parallel}{C}}} COOH \; \rightleftharpoons \; R_2 \underset{NH_2}{\overset{H}{\underset{|}{\overset{|}{C}}}} COOH \; + \; R_1 \underset{O}{\overset{}{\underset{\parallel}{C}}} COOH \qquad \text{(i)}$$

$$R \underset{NH_2}{\overset{H,,,}{\underset{|}{\overset{|}{C}}}} COOH \; \rightleftharpoons \; H \underset{NH_2}{\overset{R,,}{\underset{|}{\overset{|}{C}}}} COOH \qquad \text{(ii)}$$

$$XCH_2(CH_2)n \underset{NH_2}{\overset{H}{\underset{|}{\overset{|}{C}}}} COOH \; \rightleftharpoons \; CH_3(CH_2)n \underset{O}{\overset{}{\underset{\parallel}{C}}} COOH \; + \; XH \; + \; NH_3$$
$$(n = 0, 1) \qquad \text{(iii)}$$

$$XCH_2(CH_2)n \underset{O}{\overset{}{\underset{\parallel}{C}}} COOH \; + \; YH \; \rightleftharpoons \; YCH_2(CH_2)n \underset{O}{\overset{}{\underset{\parallel}{C}}} COOH \; + \; XH$$
$$(n = 0, 1) \qquad \text{(iv)}$$

$$R \underset{NH_2}{\overset{H}{\underset{|}{\overset{|}{C}}}} COOH \; \rightleftharpoons \; R \underset{NH_2}{\overset{H}{\underset{|}{\overset{|}{C}}}} H \; + \; CO_2 \qquad \text{(v)}$$

$$R \underset{H \; NH_2}{\overset{HO \; H}{\underset{|\;\;|}{\overset{|\;\;|}{C}}}} COOH \; \rightleftharpoons \; H \underset{NH_2}{\overset{H}{\underset{|}{\overset{|}{C}}}} COOH \; + \; RCHO \qquad \text{(vi)}$$

Eq. i-vi

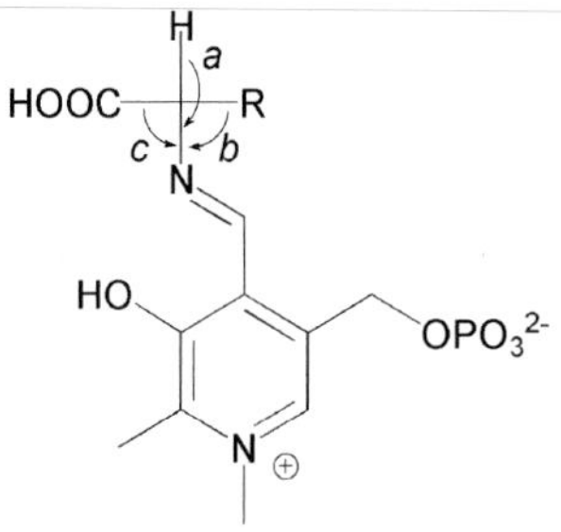

a: Transamination and racemization

b: Aldolase-type reaction

c: Decarboxylation

Scheme 2.1

an ingenious interplay between cofactor and active site organization, the required reaction is optimized while all possible others are nearly completely prevented. Vitamin B6-dependent enzymes have thus evolved to be highly efficient in terms of speed, as well as in terms of reaction and substrate specificity.

The pioneering work of Snell and Braunstein helped to clarify the mechanisms of all the PLP and PMP-dependent reactions [4]. In simple model systems, pyridoxal and pyridoxamine (with or without their phosphate group) could perform essentially all of the reactions in which they are involved metabolically. However, these non-enzymatic reactions were very slow, and the characteristic selectivities of the enzymatic processes were not present. This motivated us and several other groups to construct more sophisticated model systems to imitate vitamin B6-dependent enzymes, with several goals. The first goal is to improve the reaction rate. The second is to attain substrate selectivity, chemical reaction selectivity, regioselectivity, and stereoselectivity. A further goal is to achieve turnovers in the catalysis. Herein we review the progress made in this field [5].

2.2
Transamination

Almost all amino acids are synthesized by the biochemical reaction of PMP with α-keto acids, in one of the steps of an overall process referred to as transamination (Scheme 2.2). PMP is converted into PLP, and PLP can react with a second amino acid to regenerate PMP and convert the amino acid into its corresponding keto acid. As a result, a keto acid and an unrelated amino acid interchange functionality.

2.2.1
Pyridoxamines with Small Auxiliary Groups

A basic group of the enzyme catalyzes the proton transfer characteristics of transaminations. Bruice found that imidazole buffer can catalyze transamination reactions in model systems [6]. We conjectured that transamination rates should be improved by attaching a basic side arm to pyridoxamine. Thus we synthesized a series of simple pyridoxamine derivatives carrying basic groups at the end of flexible chains of various lengths [7–9]. We measured the transamination rates of these pyridoxamine derivatives in the presence of Zn(OAc)$_2$ at pH 4.00 in methanol.

The results (summarized in Table 2.1) showed that all the pyridoxamines with basic side chains led to significant accelerations relative to pyridoxamine itself (2) and to 1. For the N,N-diaminoalkyl series, catalyst 4 with two methylene groups carries the dimethylamino function so far out that it would form a nine-membered ring on removing the proton from the C4' carbon of the ketimine. From the work of Hine [10], this was more than long enough to permit such a process, but interestingly this was not the optimum chain. Compound 5, with one additional methylene, showed an improved rate. This makes no sense if the dimethylamino group is functioning only as a base – but this increased chain length would be needed if, after removing the proton from the

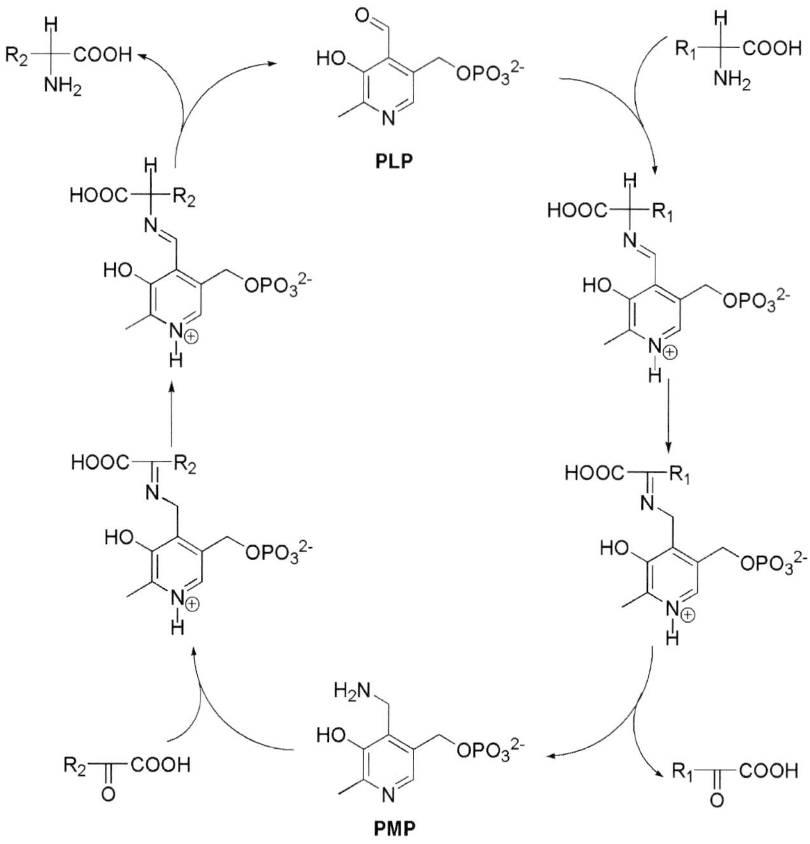

Scheme 2.2

C4′ carbon, the dimethylamino group then transferred the proton to the α-carbon of the developing amino acid. Thus this preference for increased chain length seems to argue for such a dual catalytic role of the amine. As expected from this, compound **6** with yet an additional methylene group, which is no longer needed to permit the α-carbon to be reached, is now less active.

Related but even more striking effects are seen with the imidazole catalysts (**8** and **9**). This might simply reflect the rigidity in an imidazole ring, with a resulting entropy advantage for these systems. However, imidazole is a considerably weaker base than an alkyl amine, and weaker basicity should be an advantage in this catalytic process since the base is not fully protonated at the operating pH. In addition, ethylenediamine systems **13–15** are remarkably good catalysts. They improve the transamination rate by 100–200-fold. This is because the diamine is acting as a mono-protonated bifunctional catalyst, with one amino group removing the proton from the C4′ carbon while the other puts a proton on the α carbon of the forming amino acid. Molecular models show that this bifunctional mechanism is possible for **13–15** but not for **12**, which gives a much slower rate.

Table 2.1 Relative transamination rates at pH 4.0 in methanol with Zn(OAc)$_2$ of small pyridoxamine analogs incorporating basic side arms.

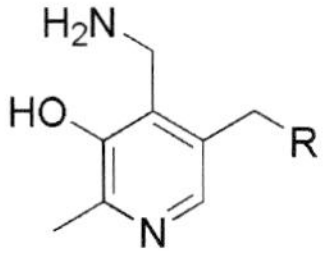

Compound	R	k_{rel}
1	SCH$_2$CH$_2$CH$_3$	1.00
2	OH	1.4
3	NMe$_2$	14
4	SCH$_2$CH$_2$NMe$_2$	26
5	SCH$_2$CH$_2$CH$_2$NMe$_2$	37
6	SCH$_2$CH$_2$CH$_2$CH$_2$NMe$_2$	13
7	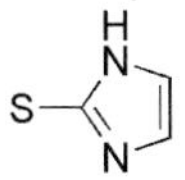	5.6
8	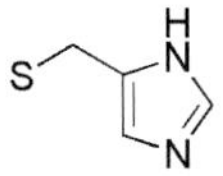	12
9	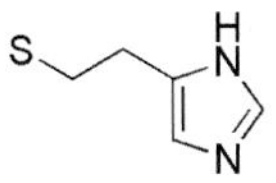	76
10	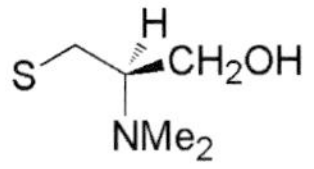	26
11	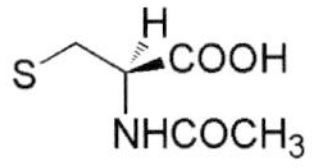	11
12	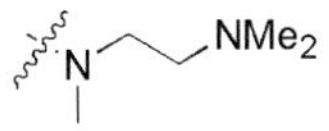	27
13	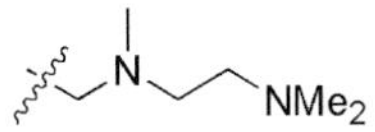	110
14	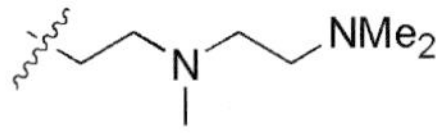	87
15	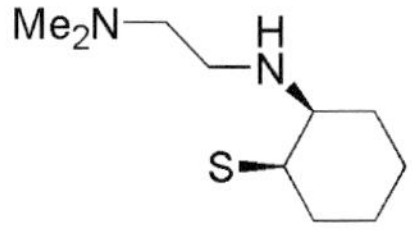	200

With the clear evidence that a basic side arm attached to pyridoxamine could act not only to remove a proton but also subsequently to add it to the α-carbon of the developing amino acid, it seemed attractive to learn how to perform this reaction with high stereoselectivity. Attachment of a chiral side chain to pyridoxamine could induce some stereoselectivity, but the enantiomeric excess (ee%) in these cases were usually low. For instance, compound **10** with an *N,N*-dimethylcysteinol side chain produced D-norvaline from α-ketovaleric acid in 39% ee, while **11**, carrying an *N*-acetylcysteine side chain, produces a nearly racemic mixture of amino acids [7]. For stereoselective synthesis of a product amino acid the catalytic side arm must operate on one face or the other of this planar system. This could be achieved by fusing a ring to the pyridoxamine unit, so that a side arm coming off this ring would necessarily be aimed either forward or backward.

Thus we designed and synthesized a bicyclic pyridoxamine derivative carrying an oriented catalytic side arm (**16**) [11]. Rates for conversion of the ketimine Schiff base into the aldimine, formed with **26** (below) and α-ketovaleric acid, indolepyruvic acid, or pyruvic acid, were enhanced 20–30 times relative to those carried out in the presence of the corresponding pyridoxamine derivatives without the catalytic side arm. With α-ketovaleric acid, **16** underwent transamination to afford D-norvaline with 90% ee. The formation of tryptophan and alanine from indolepyruvic acid and pyruvic acid, respectively, showed a similar preference. A control compound (**17**), with a propylthio group at the same stereochemical position as the aminothiol side arm in **16**, produced a 1.5:1 excess of L-norvaline, in contrast to the large preference for D-amino acids with **16**. Therefore, extremely preferential protonation seems to take place on the *si* face when the catalytic side arm is present as in **16**.

Bernauer et al. [12] developed an alternative method to induce chirality in the pyridoxamine-mediated transamination reactions. They synthesized some chiral terdentate ligands (**18–20**) having C_2 symmetry. Using these auxiliary ligands they studied the isomerization of the Cu(II) Schiff-base complex of pyridoxamine and various keto acids. In a weakly acidic solution, optically active α-amino acids with (*R*)-configuration were formed preferentially, and the maximum ee%s were 80, 54, 48, and 29 for phenylalanine, alanine, leucine, and valine, respectively (Scheme 2.3). The authors proposed that the enantioselectivity comes from formation of an intermediate Cu(II)-ketimine complex with the auxiliary ligand. However, the relatively high ee%s were only obtained at the initial reaction stage because an undesirable racemization reaction occurred subsequently.

16　　　　　　　**17**

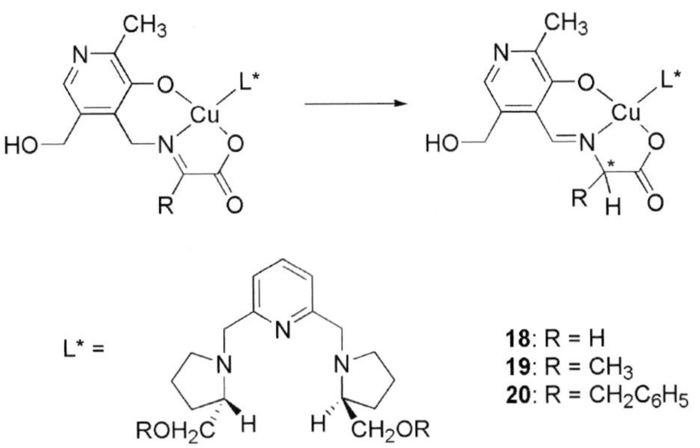

L* =

18: R = H
19: R = CH$_3$
20: R = CH$_2$C$_6$H$_5$

Scheme 2.3

A further method to induce chirality in the pyridoxamine-mediated transamination reactions was developed by Kuzuhara et al. [13]. They synthesized optically resolved pyridinophanes (**21**, **22**) having a nonbranched "ansa chain" between the 2'- and 5'-positions of pyridoxamine. With the five-carbon chain in **21** and **22**, the two isomers do not interconvert readily. In the presence of zinc(II) in organic solvents such as metha-nol, *tert*-butanol, acetonitrile, and nitromethane, they observed stereoselective transa-mination between pyridinophanes and keto acids. The highest ee%s are 95% for D- and L-leucine by reaction of the corresponding α-keto acid with (S)- and (R)-**22**, respec-tively. On the basis of kinetic analysis of the transamination reactions, Kuzuhara et al. originally proposed a mechanism for the asymmetric induction through kinetically controlled stereoselective protonation to the carbanion attached to an octahedral Zn(II) chelate intermediate. However, they subsequently raised some questions about this proposal [14].

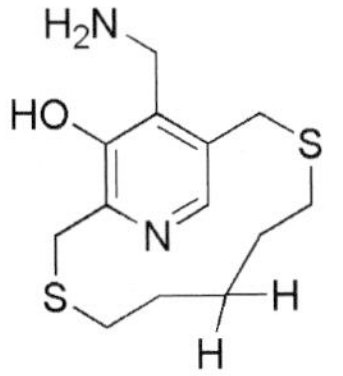

21

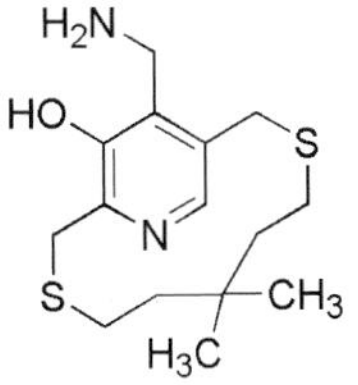

22

2.2.2
Pyridoxamines with a Binding Site – Artificial Enzymes

An important feature of the enzymatic systems is the presence of a binding site. Thus it is attractive to learn how to construct vitamin B6-dependent enzyme models that can provide a substrate binding site and perform molecular recognition. The first example was a catalyst (**23**) in which pyridoxamine was linked to the primary face of β-cyclo-dextrin (β-CD) through a sulfur atom [15]. Catalyst **23** could transform α-keto acids into α-amino acids, as pyridoxamine does, but with selectivity. That is, phenylpyruvic acid was transaminated ca. 100 times as rapidly as was pyruvic acid by **23**, while simple pyridoxamine shows no such selectivity. Compound **23** is selective because of binding of the phenyl group into the β-CD cavity. As expected from this, a better binding *p-tert*-butylphenyl group has an even larger effect: from competition studies, the preference for 4-*tert*-butylphenylalanine formation from its keto acid exceeds 15 000 times that for the amination of pyruvic acid [16].

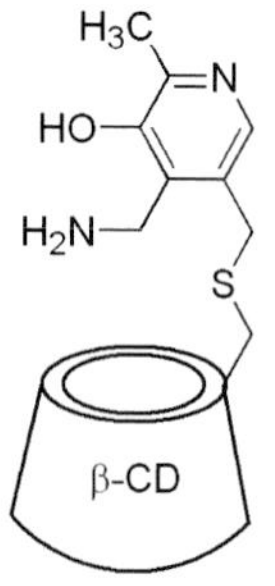

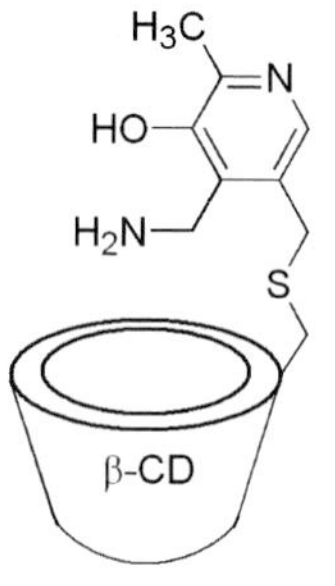

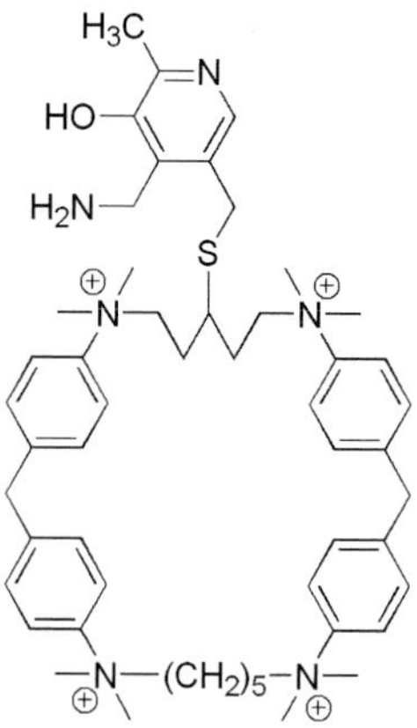

Another pyridoxamine analog (**24**), prepared with the pyridoxamine attached to the secondary face of β-CD [17], showed similar properties to those of the primary derivative. As another comparison, an artificial macrocyclic binding group (i.e. tetraaza[1.7.1.7]paracyclophane ring) was used in place of β-CD [18]. The resulting catalyst (**25**) accelerated the conversion of aromatic α-keto acids bearing good hydrophobic segments into the corresponding α-amino acids by more than one order of magnitude as compared with the rates for reactions catalyzed by pyridoxamine. To fix pyridoxamine to the cyclodextrin with even better defined geometry, a set of four compounds [**26-(1–4)**] was made by reacting a pyridoxaminedithiol with β-CD-6A,6B-diiodide [16]. As molecular models suggested, isomers **26-(1)** and **26-(2)**, with the pyridoxamine held over the β-CD cavity, preferred para-substituted phenylpyruvic acids, while the other pair held the pyridoxamine to the side and preferred meta-substituted phenylpyruvic acid substrates.

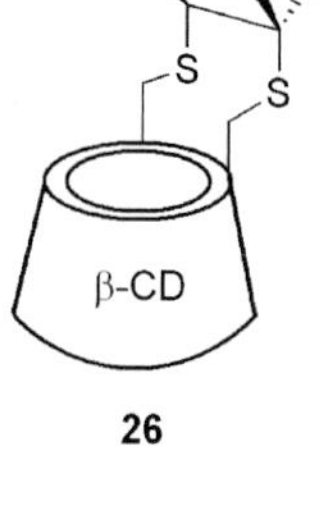

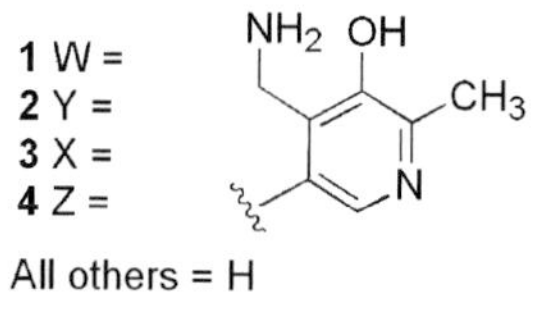

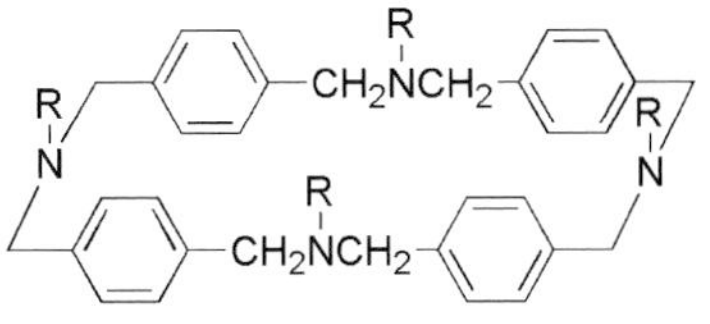

R = structure with $C_{13}H_{27}$ groups

27

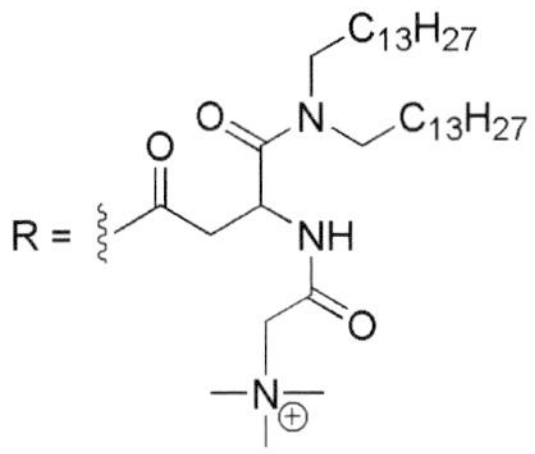

Murakami et al. [19] developed an additional vitamin B6 model system with a binding site. They synthesized an octopus cyclophane (**27**) as a functional model of the protein matrix of transaminase. This cyclophane formed a hydrophobic cavity in water where PLP could be noncovalently incorporated. Alkylamines having various hydrophobic chains were employed as substrates, in place of α-amino acids, to evaluate the hydrophobic effect on the Schiff base-forming equilibrium. The Schiff base formation constant was found to depend markedly on the chain length of a substrate in the presence of **27**, indicating that the octopus cyclophane can be utilized as an effective holoenzyme model capable of forming a ternary complex.

With the binding group in hand, it is interesting to learn how to induce stereoselectivity in the transamination. The earliest example (**23**) afforded amino acids with some stereoselectivity, because of the chirality of the cyclodextrin unit [15]. Tryptophan and phenylalanine formed by the reaction with **23** showed preferable formation of the L-enantiomers at L/D ratios of 2:1 and 5:1, respectively. More selectivity is expected if the proton is delivered by a chirally mounted basic group, as in the enzyme. Thus, compound **28**, carrying both a pyridoxamine and an ethylenediamine unit attached to β-CD on neighboring primary methylene groups, was prepared and studied for its ability to form amino acids from keto acids with chiral selectivity [20]. Although Tabushi et al. reported quite good selectivities, it has proven impossible in our laboratory to duplicate these findings. However, in some alternate approaches we have produced optical induction with related catalysts (**29**), but not in high 90% selectivities [21].

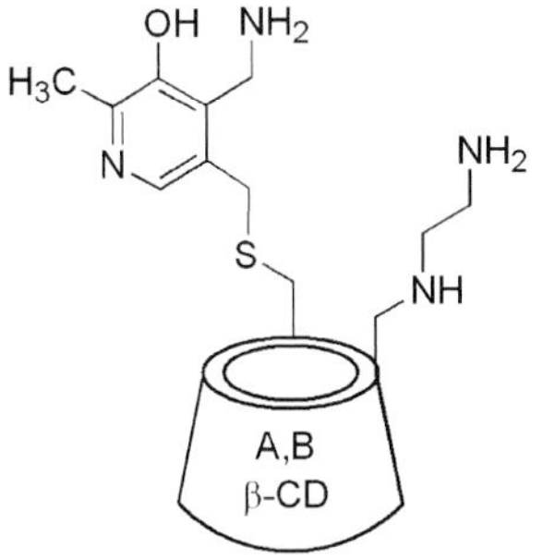

28

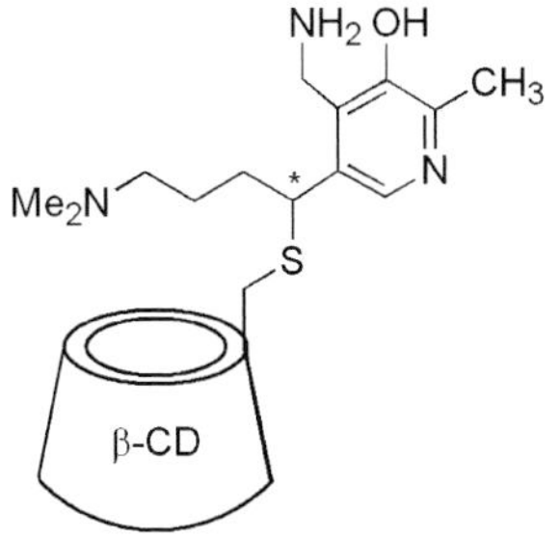

29

2.2.3
Pyridoxamine–Surfactant Systems

Surfactants provide interesting systems in bioorganic chemistry. The specific micro-environments produced by surfactant micelles in aqueous media are effective for simulation of catalytic functions in the hydrophobic active sites of enzymes. This has inspired several groups to study various pyridoxamine–surfactant systems for transaminase enzyme mimics. The first work was reported by Kondo et al. [22], who found that N-dodecylpyridoxal chloride forms a Schiff base with hydrophobic α-amino acids, such as tryptophan, phenylalanine, phenylglycine, and S-benzylcysteine, in the cationic micelle of hexadecyltrimethylammonium chloride. The formation constants were large, and the resulting Schiff bases underwent transamination to give the corresponding α-keto acids effectively, even in the absence of metal ions in aqueous media at 30.0 °C.

Murakami et al. studied alternative pyridoxamine–surfactant systems [23]. These authors synthesized hydrophobic pyridoxamine derivatives (**30** and **31**) and peptide lipid molecules (**32–35**). Catalyst **30** or **31** and the peptide lipids formed bilayer membranes in water, which showed transamination reactivity in the presence of metal ions such as Cu(II). It was proposed that the pyridoxamine moiety was placed in the so-called hydrogen-belt domain interposed between the polar surface region and the hydrophobic domain that is composed of double-chain segments within the bilayer assembly. The basic group (such as imidazole) in the peptide lipid molecules could catalyze the proton transfer involved in the transamination reaction. In addition, marked substrate discrimination by these bilayer membrane systems was performed through hydrophobic interactions between substrates and the catalytic site.

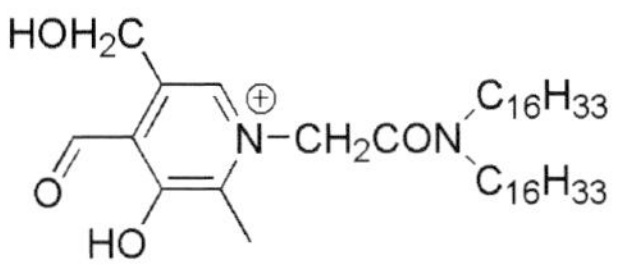

36 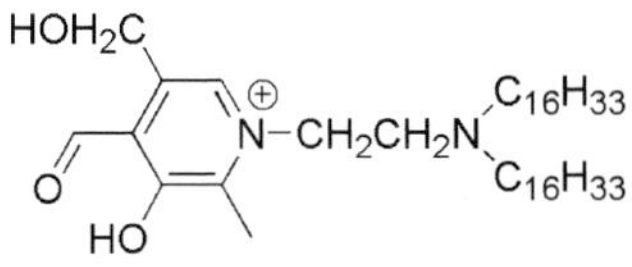   **37**

Murakami et al. also found that the transamination reaction between hydrophobic pyridoxals (**36** and **37**) and α-amino acids, to produce α-keto acids, was extremely slow for neutral pyridoxals even in the presence of Cu(II) ions [24]. Detailed kinetic analysis of the reactions carried out in the vesicular system indicated that the transformation of the Cu(II)-quinonoid chelate into the Cu(II)-ketimine chelate was kinetically unfavorable compared with the competing formation of the Cu(II)-aldimine chelate from the same quinonoid species. This problem was solved to a certain extent by quaternization of the pyridyl nitrogen in pyridoxal, as Murakami et al. successfully accomplished transamination between catalyst **36** and L-phenylalanine to produce phenylpyruvic acid.

Having successfully accelerated the reversible isomerization between the aldimine and ketimine Schiff bases, Murakami et al. then studied how to obtain turnovers in the full transamination reaction between one amino acid and one keto acid [25]. They found that the bilayer vesicle system constituted with **33**, **36**, and Cu(II) ions showed some turnovers for the transamination between L-phenylalanine and pyruvic acid. However, such turnover behavior was not observed in a vesicular system composed of **32**, **36**, and Cu(II) ions, and an aqueous system involving *N*-methylpyridoxal and Cu(II) ions without amphiphiles. Therefore, both the hydrophobic effect and the imidazole catalysis effect were proposed as important for the turnover behavior.

Murakami et al. also examined the enantioselectivity of the catalyzed transamination reaction in a bilayer membrane [26]. They contrasted a system composed of a peptide lipid bearing an L-lysine residue (**34**), a hydrophobic pyridoxal derivative quaternized at the pyridyl nitrogen (**37**), and Cu(II) ions. This system exhibited turnover behavior for

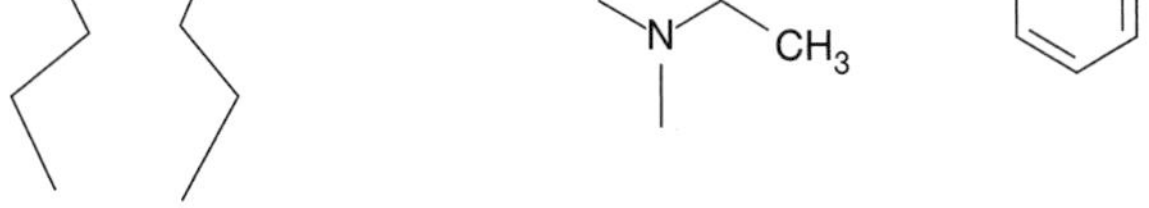

Figure 2.2 Proposed mechanism for enantioselective transamination reaction catalyzed by Murakami's bilayer membrane systems. (Reprinted with permission from Ref. 27. Copyright 1995 American Chemical Society.)

the transamination between L-phenylalanine and pyruvic acid to form D-alanine with an ee as high as 90%. Since the presence of L-phenylalanine proved indispensable for the enantioselective transamination, a five-coordinated Cu(II) complex was proposed as a key intermediate in the stereoselective protonation (Figure 2.2), in which the L-phenylalanine molecule present in excess over **37** acts as a chiral bidentate ligand to be bound to the quinoid-Cu(II) complex [27]. The ammonium group of the lysine residue of **34** acts as a proton source to protonate the quinoid-Cu(II) complex preferentially from the less hindered *si* face of the imino carbon to afford the aldimine chelate.

2.2.4
Vitamin B6–Polypeptide Systems

All the above vitamin B6 enzyme models were built with small molecules. Natural enzymes, however, are macromolecules. Their macromolecular structures offer ideal frames for the construction of versatile, robust catalytic sites. Strong and selective binding of the substrate is attained through a combination of the hydrophobic effect and specific substrate–enzyme interactions such as hydrogen bonding. The macromolecular structure can also create regions in which the catalyzed reactions occur in a less than fully aqueous medium. It is interesting to learn how to construct macromolecular enzyme mimics. A straightforward idea would be to employ polypeptides.

Imperiali and Roy were the first to introduce a pyridoxal cofactor to a polypeptide [28–30]. Their idea was to incorporate a pyridoxal-based amino acid residue in the S-peptide of the RNase-S complexes. These RNase-S complexes are semisynthetic constructs consisting residues 21–124 of native ribonuclease A (S-protein) and an oligopeptide based on residues 1–20 (S-peptide) (Figure 2.3). Analysis of the crystal structure of RNase-S revealed that a $Phe_8 \rightarrow$ pyridoxal (Pal_8) substitution in the S-peptide would position the residue in proximity to the general acid–base pair (His_{12}, His_{119}) utilized by the native enzyme. Thus, a few S-peptides containing a Pal_8 residue were designed, synthesized by standard Fmoc solid phase procedures. These Pal_8-containing S-peptides bound with the S-protein with affinities comparable to that of native S-peptide. The S-peptide-S-protein complexes were then evaluated for their ability to convert L-alanine into pyruvate under single turnover conditions in the presence or absence of copper(II). In one case, the binding to S-protein increased the transamination rate of the Pal_8-containing S-peptide by 16.7-fold.

In a similar approach, pyridoxamine was introduced into an S-peptide at position 8 to maintain the interactions with His_{12} and His_{119} [30]. Upon formation of the RNase complex, the rate was enhanced 7-fold compared with uncomplexed peptides under single turnover conditions. However, replacing the His residue at position 12 with Ser afforded only a 3-fold rate increase for the S-peptide-S-protein complex. Under catalytic conditions with pyruvate and L-phenylalanine as the substrates, uncomplexed peptides did not show catalytic turnover, suggesting that a hydrophobic microenvironment in the peptide–protein complex is critical for catalysis. However, in the presence of the S-protein, catalysis ensued. Up to 1.5 turnovers were observed in 160 h from the S-peptide-S-protein complex.

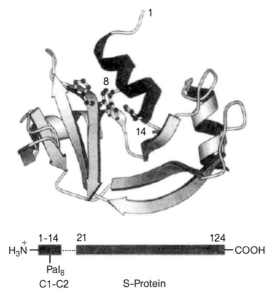

Figure 2.3 Proposed model for the semisynthetic RNAse-S complex incorporating a pyridoxal-based amino acid residue. The S-peptide fragment is highlighted, illustrating the location of pyridoxal (Pal$_8$) and the general acid-base pair. (Reprinted with permission from Ref. 28. Copyright 1994 American Chemical Society.)

In an alternative approach to generating the polypeptide-vitamin B6 system Distefano and co-workers [31] employed a family of small lipid-binding proteins to which pyridoxamine could be covalently attached through a disulfide linkage. Initially they utilized the adipocyte lipid-binding protein (ALBP, Figure 2.4), which has a cysteine residue (Cys$_{117}$) [32]. After attachment of pyridoxamine, the resulting ALBP–pyridoxamine conjugate was able to reductively aminate various α-keto acids to amino acids, with enantioselectivities ranging from 0 to 94 % ee. The reaction rates of these reactions were not, however, significantly faster than those involving free pyridoxamine. This suggested that the protein cavity was functioning as a chiral environment that controls the facial selectivity of the protonation of the aldimine intermediate without forming specific interactions with the bound pyridoxamine cofactor to accelerate the reaction.

In subsequent studies, Distefano et al. utilized a different fatty acid-binding protein, i.e. intestinal fatty acid-binding protein (IFABP) [33]. The wild-type IFABP has no Cys inside the cavity, and thus it is an ideal template to introduce single Cys residues at different positions. Cys mutations at various positions were introduced using site-directed mutagenesis. By varying the position of the cofactor attachment, and thereby changing the microenvironment of the "active site", the rate, enantioselectivity, and substrate selectivity of the protein–pyridoxamine conjugate were modulated.

Further studies on these protein–pyridoxamine conjugates were devoted to the turnover behaviors, metal ion effects, and effects of pyridine quaternization [34]. As many

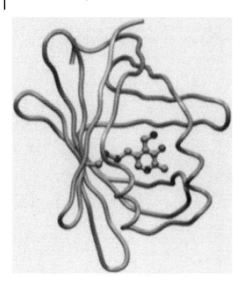

Figure 2.4 Stereoview of a model for the conjugate ALPB-pyridoxamine, showing pyridoxamine attached to a site in the protein interior. (Reprinted with permission from Ref. 32. Copyright 1996 American Chemical Society.)

as 50 turnovers were observed for the transamination between phenylalanine and α-ketoglutarate, producing L-glutamate with an enantiomeric purity of 93 % ee. The rate was enhanced by as much as 4200-fold with the protein–pyridoxamine conjugate when compared with the unliganded pyridoxamine phosphate. Although these results are still quite primitive compared with natural enzyme systems, they clearly show progress in the right direction.

2.2.5
Polymeric and Dendrimeric Vitamin B6 Mimics

We recently utilized synthetic polymers and dendrimers to construct macromolecular transaminase mimics. Initially, we covalently attached pyridoxamine to commercial polyethylenimine (PEI) [35]. The PEI polymer is highly branched, has a molecular number-average (M_n) weight of about 60 000, and contains about 1400 monomer residues. About 25 % of the amino groups of PEI are primary, about 50% secondary, and the remaining are tertiary. To imitate the hydrophobic environment provided by the enzymes, we covalently attached some alkyl groups to the PEI–pyridoxamine reagent (Figure 2.5).

The PEI–pyridoxamine polymer was treated with excess pyruvic acid in various buffer solutions, without added metal ions and with excess added EDTA. Kinetic studies revealed that the attached polymer increased the rate of pyridoxamine transamination with pyruvic acid by a factor of 6700–8300 at pH 5.0. Under higher pH conditions, the rate enhancement decreased: At pH 7.0 the rate enhancement by the polymer was still 2300 times, while at pH 8.0, the optimum for pyridoxamine itself, it was 1900. We also found that transamination by simple pyridoxamine showed strong metal ion catalysis – adding 1 equiv. of $CuCl_2$ per pyridoxamine unit to the pH 5.0 solution increased the

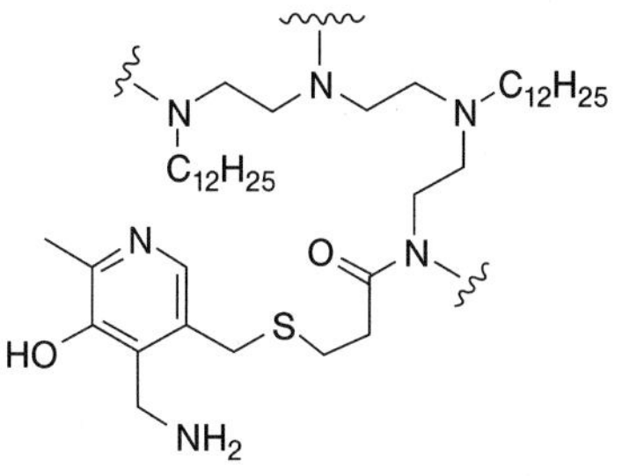

Figure 2.5 Structure of the PEI-pyridoxamine systems. (Reprinted from Ref. 36. Copyright 2002 American Chemical Society.)

free pyridoxamine rate constant by 6000 times. However, the rate for the polymer increased by only 30-fold. The transaminase enzymes do not use metal ions.

The rate enhancement of the polymer over that of simple pyridoxamine was a steep function of the length of the alkyl chains added, in polymers with roughly the same percentage of alkylation and of pyridoxamine attachment. At pH 7.0 and 30 °C, the acceleration over the rate with pyridoxamine was 160 for C-1 chains, 180 for C-3, 500 for C-6, 1000 for C-9, 2300 for C-12, and 2500 for the C-15 and C-18 normal alkyl chains. This chain effect seems unlikely to involve hydrophobic binding of a substrate as hydrophilic as pyruvic acid. Instead the hydrophobic chains modify the pK_as of the amino groups in the polymer and also create a cavity in which the transamination can take place in a less than fully aqueous environment.

Our further study on the PEI–pyridoxamine systems revealed that the amination reaction of pyruvic acid with simple pyridoxamine was buffer catalyzed, but the reaction with our polymer-linked pyridoxamine showed no catalysis by external buffers [36]. This supported the argument that the acid and base groups of the polymer were performing catalytic proton transfers in the transamination process. Extrapolating [buffer] to zero, the PEI–pyridoxamine system was actually 10 000-fold faster than was simple pyridoxamine with pyruvic acid at pH 7.0. Furthermore, the polymer system showed saturation effects when titrated with substrates (Michaelis–Menten kinetics). Laurylation of the polymers both increased k_2 by producing a less aqueous reaction environment and selectively bound the more hydrophobic substrates, decreasing K_M. This resulted in a high rate acceleration for all substrates, and a particularly high acceleration for hydrophobic substrates. The amination rate of indolepyruvic acid to tryptophan was enhanced 240 000-fold relative to that with simple pyridoxamine.

To study the effect of polymer size on catalysis [37], pyridoxamine was linked to a series of PEIs with M_n = 600, 1800, 10 000, and 60 000, both simply permethylated and with additional attached dodecyl chains. The polymers were examined in the transamination of pyruvic acid and of phenylpyruvic acid, showing Michaelis–Menten behavior. The k_2 and of K_M determined showed only small variations with polymer size. Thus, the strong advantage of pyridoxamines attached to the M_n = 60 000 PEI, relative to simple pyridoxamine alone, was seen to almost the same extent with the smaller PEIs.

Notably, the PEI enzyme models do not have a well-defined structure, and each polymer molecule contains more than one coenzyme unit. To make more defined macromolecular transaminase models we synthesized several water-soluble poly(amidoa-

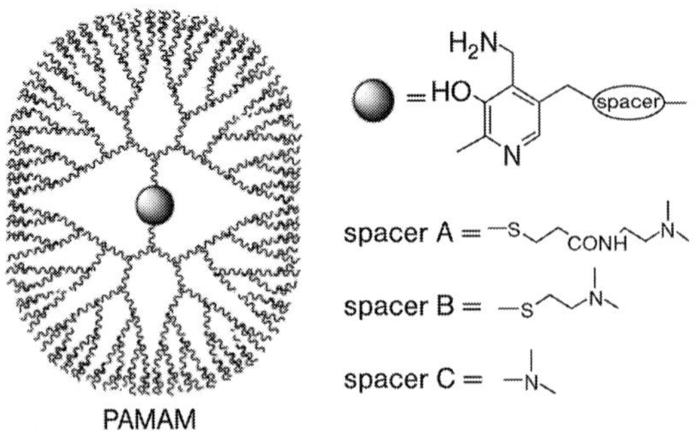

Figure 2.6 Structure of the PAMAM-pyridoxamine systems. (Reprinted from Ref. 38. Copyright 2003 American Chemical Society.)

mine) (PAMAM) dendrimers in which we located one pyridoxamine unit at the core (Figure 2.6) [38]. Unlike PEIs, whose structures are poorly defined, PAMAM dendrimers are good mimics of globular proteins. We tuned the structure and size of the dendrimeric enzyme models, which greatly affected the transamination rates. We established that a single pyridoxamine in the core of a PAMAM dendrimer is comparable to a pyridoxamine unit linked to $M_n = 60\,000$ PEI in transaminating pyruvic and phenylpyruvic acids to alanine and phenylalanine, but it is less effective than the laurylated PEI–pyridoxamine.

2.3
Racemization

Amino acid racemases are important for bacteria because they need D-alanine in the biosynthesis of cell walls. These enzymes require pyridoxal as the active cofactor. A racemization reaction starts with the aldimine complex between pyridoxal and an α-amino acid (Scheme 2.4). Deprotonation occurs at the α-carbon of amino acid, due to the electron-sink effect of pyridoxal. Reprotonation of the quinonoid intermediate at the opposite side provides the desired product (pathway *a* in Scheme 2.4). However, reprotonation may also take place at the C4′ of pyridoxal (pathway *b* in Scheme 2.4). This kills the catalyst because one of its product, pyridoxamine, can no longer racemize an amino acid.

Thus, to attain the racemase reactivity it is important to learn how to block the transamination process. Our idea was to place a catalytic group that can protonate only the α-carbon of the amino acid unit but not the C4′ of the pyridoxal moiety in the quinonoid intermediate (Scheme 2.5). Thus we synthesized catalyst **38**, which carries a rigid pyridine side chain [39]. Catalyst **39**, which lacks the double bond, was also synthesized as a less rigid control. Both catalysts catalyzed loss of optical activity from the aldimine equally well – about twice as fast as simple pyridoxal. However, **39** could catalyze the transamination reaction 2.5 times faster than **38**. Therefore, **38** showed a small preference for racemization over the transamination reaction as compared with **39** by a

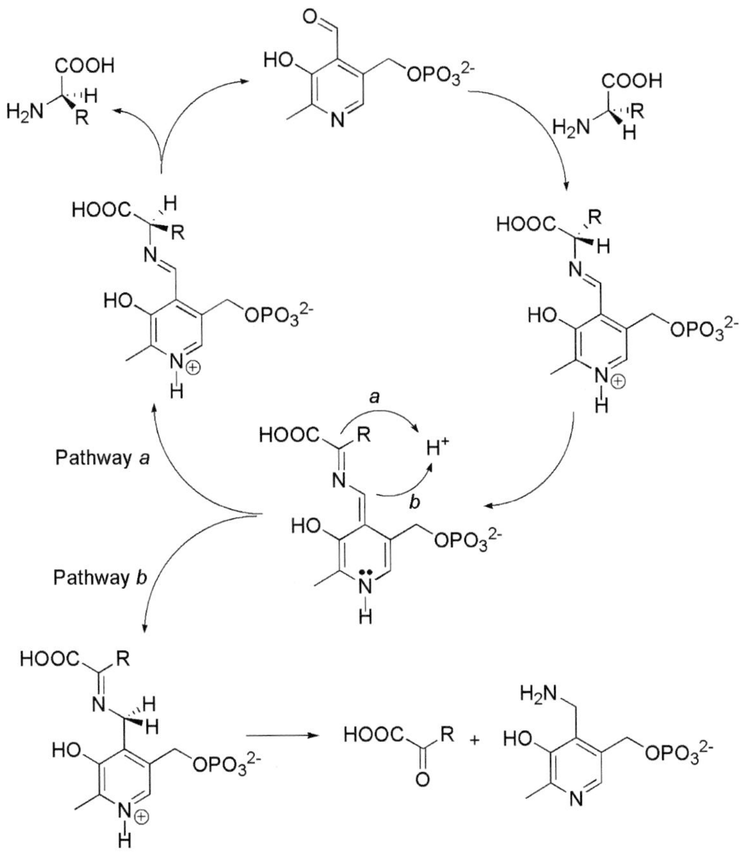

Scheme 2.4

matter of two-fold or so. In our second generation of pyridoxal catalysts for racemization, we used a benzene ring as the rigid spacer [40]. Compound **40** can selectively catalyze racemization of an amino acid in preference to transamination about 11 times better than does the control compound **41**.

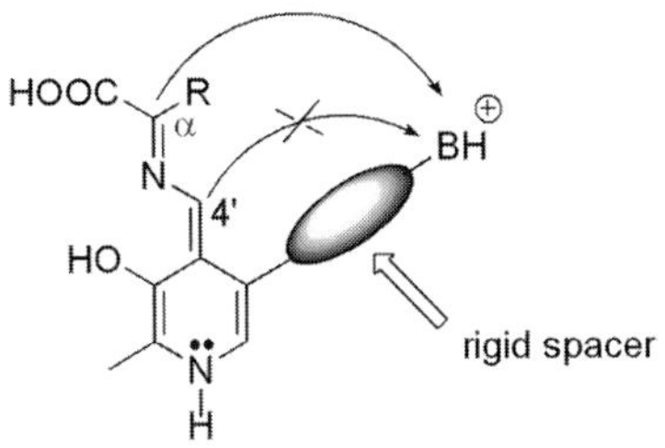

Scheme 2.5

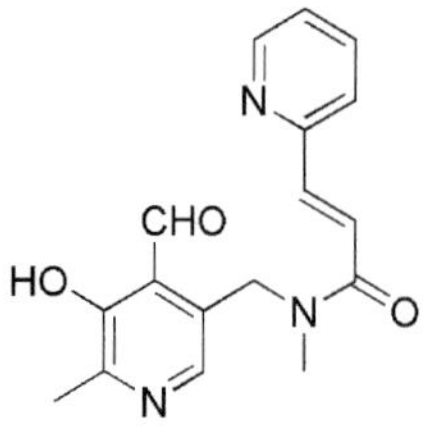

38

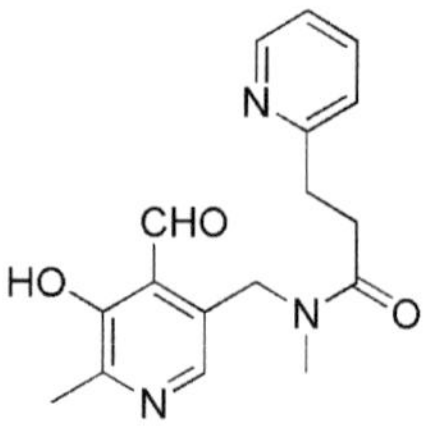

39

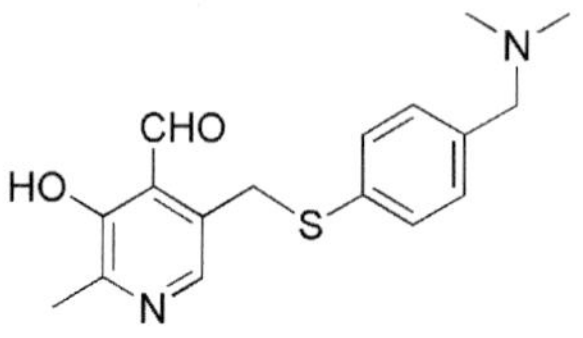

40

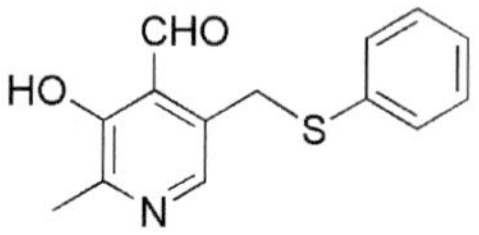

41

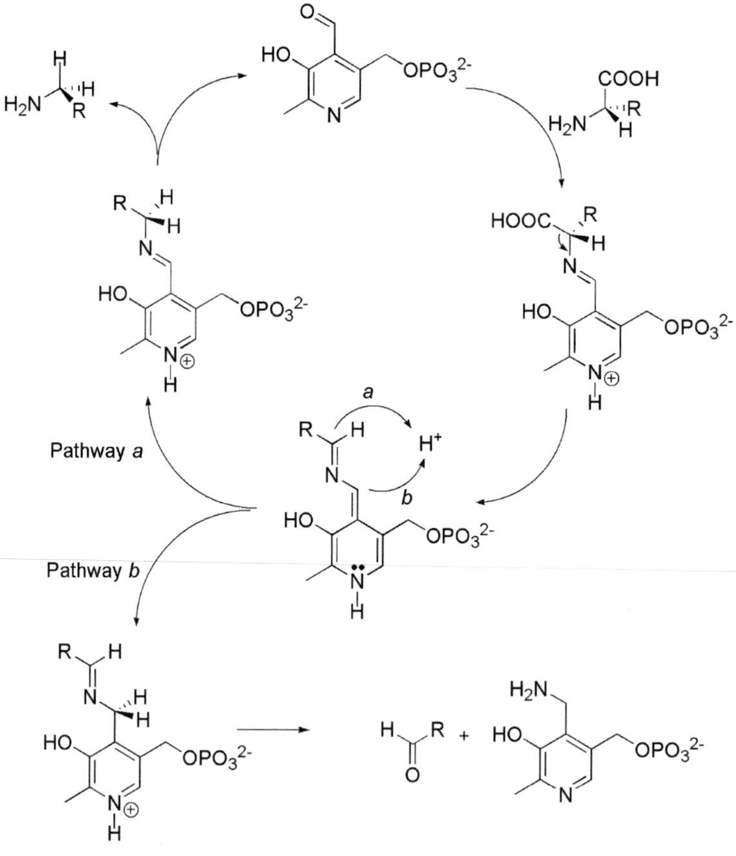

Pathway *a*

Pathway *b*

Scheme 2.6

2.4
Decarboxylation

Decarboxylation of an amino acid is an important reaction, catalyzed by a pyridoxal-dependent decarboxylase, that affords an amine as product (Scheme 2.6). It is very attractive to learn how to mimic this process to generate various amines from α-amino acids. Unfortunately, our previous studies established that treatment of α-alkyl amino acids with pyridoxal afforded only ketone and pyridoxamine as products, by a transamination-dependent oxidative decarboxylation process (pathway *b* in Scheme 2.5) [41]. Consequently, non-oxidative decarboxylation, using pyridoxals to generate amines, remains elusive.

We also studied the pyridoxal-dependent decarboxylation of an aminomalonic acid, a process providing an α-amino acid as the product. Our interest was to induce stereoselectivity in the process. Therefore, we synthesized catalysts **42** and **43**, which rigidly held chirally mounted groups [42]. With the basic **42** we obtained 42 % ee favoring L-phenylalanine in the decarboxylation of 2-amino-2-benzyl-malonic acid, while with non-basic **43** the ee was too low to detect. We proposed that the basic side chain delivered a proton to the decarboxylation intermediate in a stereoselective fashion.

42   **43**

2.5
β-Elimination and Replacement

β-Elimination and then replacement reaction of an α-amino acid with a nucleophile is very attractive from the viewpoint of synthetic organic chemistry because various β-substituted alanines may be prepared from a simple α-amino acid, such as serine, and nucleophiles. A reaction catalyzed by tryptophan synthase – the formation of tryptophan from serine and indole – is one of the most well-known β-elimination and replacement reactions (Scheme 2.7). Here, an aldimine Schiff base is derived from reaction of the enzyme-bound PLP with serine, which then dehydrates to give the Schiff base of PLP with 2-aminoacrylate. Indole then adds to the vinyl Schiff base, generating tryptophan after lysine aminolysis of the Schiff base product.

We studied the β-elimination and replacement reactions using our pyridoxamine analogs [43]. Initially, we examined the relative rates for the β-elimination reactions between chloropyruvic acid and small pyridoxamine analogs carrying a basic side chain (Table 2.2). In the dimethylamino series the HCl elimination rates were fastest with the shorter chain of **4**, as expected for a process that requires only proton removal from the pyridoxamine 4'-CH$_2$ group. The contrast with the data for transamination

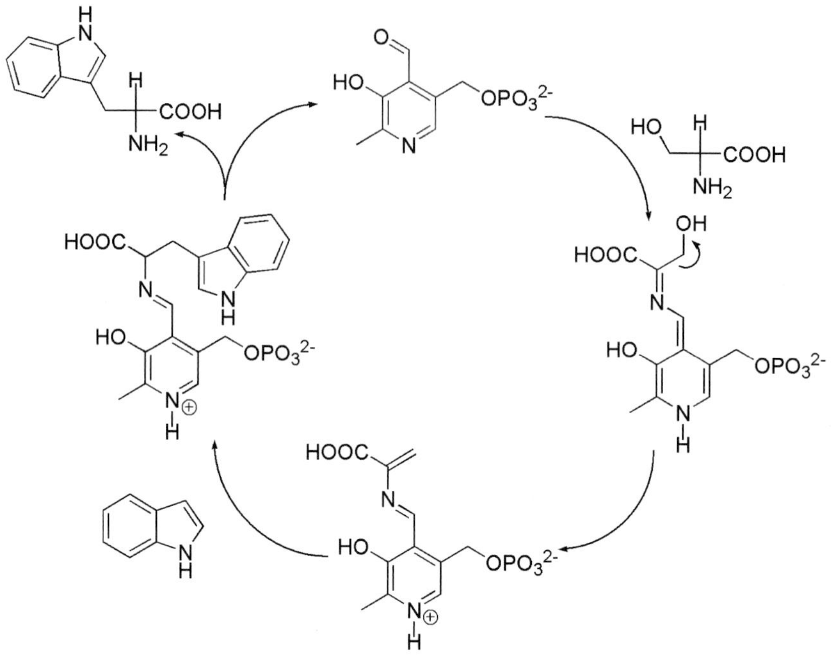

Scheme 2.7

(Table 2.1), in which **5** with the longer chain was the fastest, supports our previous contention that in transamination the catalytic group also performs the protonation at the amino acid α-carbon. In the imidazole series (**7–9**) the shortest chain system **7** was also fastest in HCl elimination. The striking contrast to the data for transamination fully supports the proposition that in our transamination studies the catalysis was sequential, with proton transfer by the catalytic group to the remote position of the intermediate.

We next synthesized the pyridoxal-bound β-CD catalyst **44** (Scheme 2.8) [43], which produced 3–5 times more tryptophan when incubated with indole, β-chloroalanine, and Al$_2$(SO$_4$)$_3$ (pH 5.2 and 100 °C) than the reaction in which the pyridoxal derivative was replaced by simple pyridoxal. However, tryptophan yield was still only a few percent. As expected, this kinetic advantage disappeared at higher indole concentrations due to saturation of the binding site. Furthermore, L-tryptophan was produced in ca. 10% excess relative to the D-enantiomer.

Murakami et al. utilized catalytic bilayer membranes to catalyze the β-replacement reaction of serine with indoles [44]. The bilayer vesicle formed with **32** and **36** drastically accelerated the β-replacement reaction by 51-fold (k_{rel}) relative to pyridoxal in homogeneous aqueous solution. They attributed this to the hydrophobic microenvironmental effect provided by the bilayer vesicle, which affords effective incorporation of indole molecules and elimination of water molecules in the reaction site. The imidazolyl group of **33** enhanced the reaction further, k_{rel} being 130, possibly due to general acid–base catalysis by the imidazolyl group. Copper(II) ions also improved the reaction.

Table 2.2 Relative rates for β-elimination reactions between small pyridoxamine analogs and chloropyruvic acid.

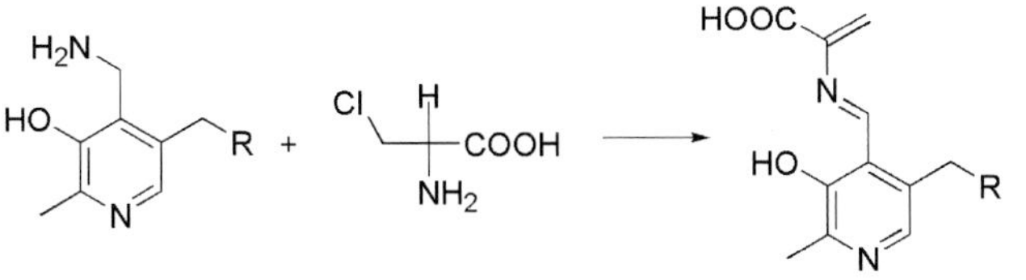

Compound	R	k_{rel}
1	$SCH_2CH_2CH_3$	1.00
3	NMe_2	4.3
4	$SCH_2CH_2NMe_2$	9.2
5	$SCH_2CH_2CH_2NMe_2$	2.6
7	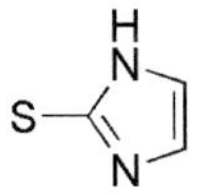	6.4
8	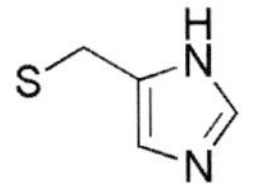	5.3
9	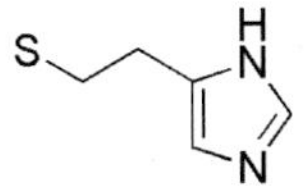	2.8

As a whole, the bilayer catalyst composed of **33**, **36**, and Cu(II) ions exhibited the highest activity for the β-replacement reaction, showing turnover behavior, with $k_{rel} = 180$.

Murakami et al. also found that, for the β-replacement reaction carried out in the **33–36** vesicular system, L-serine and its hydrophobic benzyl ester derivative show com-

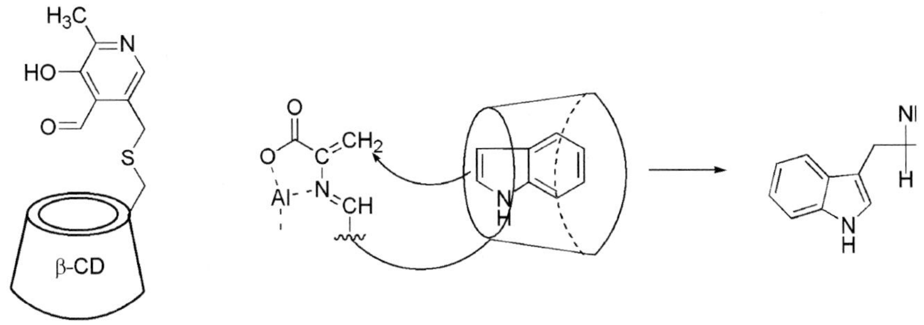

44

Scheme 2.8

parable reactivity. The formation of an aldimine Schiff base of **36** with the substrate α-amino acid was not the rate-determining step. Conversely, indole derivatives showed marked substrate specificity; relative reactivities for the reactions of L-serine with 5-hydroxyindole, 5-methylindole, 5-methoxyindole, and indole were 40, 18, 13, and 1, respectively. Such selectivity for indole derivatives primarily reflects differences in their nucleophilicity.

Modest enantioselectivity was observed for the β-replacement reaction catalyzed by **33–36** bilayer membranes [45]. D-Tryptophan formation prevailed over that of the L-form in 50–55 % ee regardless of chirality of the substrate, serine. Conversely, no enantioselectivity was observed when the **32** vesicle was used in place of **33**. This suggested that the imidazolyl group of **33** might exercise stereospecific acid catalysis in the protonation of the prochiral carboanion intermediate. The enantioselectivity was also modest (ca. 30 % ee) when the β-replacement reaction was mediated by the co-vesicle formed with **37**, **32**, and an additional peptide lipid having (*S*)-binaphthol moieties (**35**) in the presence of Cu(II) ions.

2.6
Aldolase-type Reactions

Serine hydroxymethyltransferase is a PLP-dependent aldolase. It catalyzes interconversion between glycine and various β-hydroxy-α-amino acids, such as serine and threonine, via formation of a quinoid intermediate derived from PLP with the amino acid substrate (Scheme 2.9). This aldolase-type reaction is of interest as an asymmetric synthesis of α-amino acids via C–C bond formation.

Kuzuhara et al. synthesized an optically resolved pyridoxal analog having an "ansa chain" between the 2′- and 5′-positions (**45**) [46]. The aldolase-type reaction of **45** and glycine with either acetaldehyde or propionaldehyde afforded the corresponding β-hydroxy-α-amino acid with 27–77 % ee. The erythro isomers were 1.2–1.8 times dominant over threo ones. The (*S*)-enantiomer of the pyridoxal derivative furnished the (*S*)-amino acid in excess. Accordingly, the reaction occurred on the same face as was occupied by the ansa chain. We have confirmed these results [47].

We also prepared a chiral cyclophane derivative of pyridoxal (**46**) that has amino groups oriented specifically over one face of the cofactor [47]. The (*S*)-isomer of this compound mediated the formation of threonine and *allo*-threonine from glycine and acetaldehyde with enantioselectivities that were a function of pH, reversing the chiral selectivity from low to high pH. Explanations were advanced for this stereochemical reversal and for the otherwise surprising preference of the former compounds to react on the more hindered face of the pyridoxal. We suggested that the reaction intermediate was geometrically distorted by the transannular chain, leading to reaction on the face that carried the chain due to a stereoelectronic effect (Figure 2.7). The stereochemical reversal seen with **46**, as controlled by pH, may reflect catalysis by the protonated form (Figure 2.8), but a coordination effect of metal ions bound to the protonation-free form cannot be excluded. We also recently observed a similar stereochemical reversal with catalyst **42** [42].

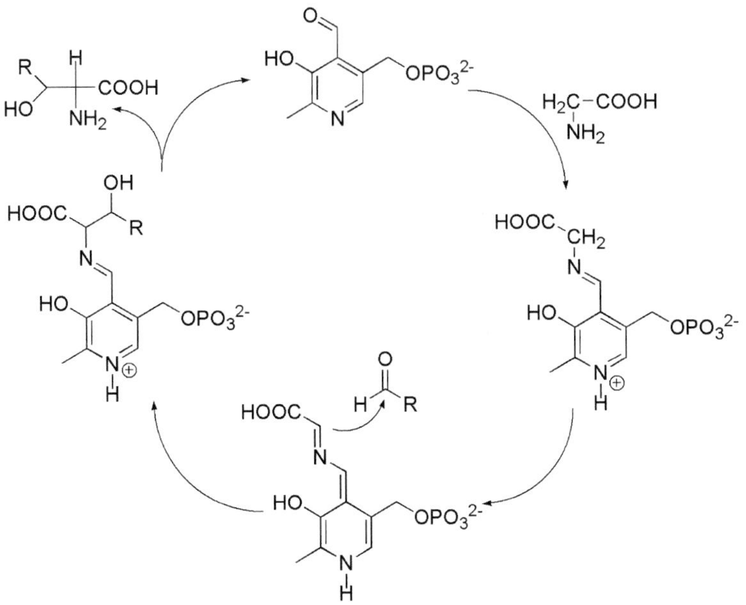

Scheme 2.9

Murakami et al. have utilized bilayer vesicles to study aldolase-type reactions [48]. Formation of β-phenylserine from glycine and benzaldehyde proceeded effectively by cooperative catalysis of a hydrophobic pyridoxal derivative (**47**) and Zn(II) ions in the bilayer vesicle formed with **32**. The threo isomer was dominantly produced over the erythro form. A marked enantioselectivity was observed in the co-vesicle of **32** and **35** in combination with **47** and Cu(II); the ee for formation of (2*S*,3*R*)-β-phenylserine over its enantiomeric (2*R*,3*S*)-isomer was 58%. Enantioselectivity also arose with another bilayer assembly, formed with **32**, **35**, and **37** in the presence of Cu(II), where the (2*R*,3*S*) isomer was dominant over the (2*S*,3*R*) species in 13% ee. The opposite enantioselectivity performed by the second system, as compared with that for **47**, might reflect a different stereochemical environment around the quinoid intermediate that allows the attack of benzaldehyde.

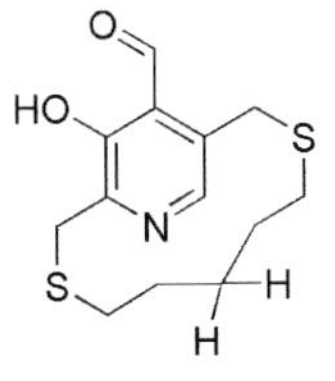

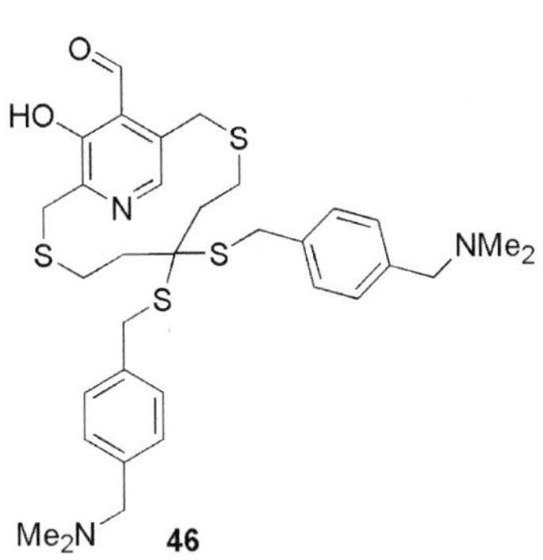

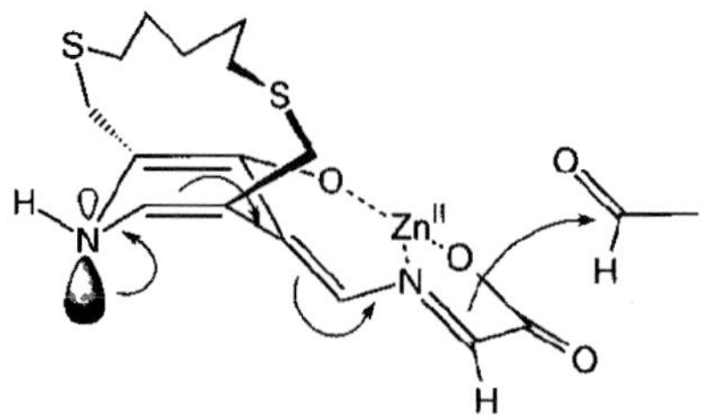

Figure 2.7 Stereoelectronic explanation of the preferential reaction of many pyridoxal ansa compounds on their seemingly more hindered faces. (Reprinted from Ref. 47. Copyright 1994 American Chemical Society.)

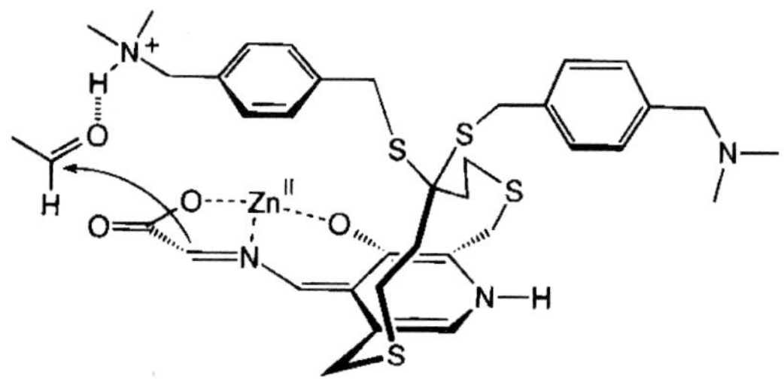

Figure 2.8 Catalysis by the protonated amine could explain the stereochemistry of threonine and *alo*-threonine synthesis by our catalyst **46** at low pH. (Reprinted from Ref. 47. Copyright 1994 American Chemical Society.)

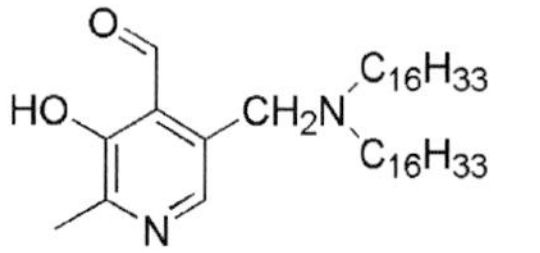

47

2.7
Conclusion

Vitamin B6 enzyme models that can catalyze five types of reactions – transamination, racemization, decarboxylation, β-elimination and replacement, and aldolase-type reactions – have been reviewed. There are also five approaches to construct the vitamin B6 enzyme models: (i) vitamin B6 augmented with basic or chiral auxiliary functional groups; (ii) vitamin B6 having an artificial binding site; (iii) vitamin B6–surfactant systems; (iv) vitamin B6–polypeptide systems; (v) polymeric and dendrimeric vitamin B6 systems. These model systems show rate enhancement and some selectivity in vitamin B6-dependent reactions, but they are still primitive compared with the real enzymes. We expect to see improved reaction rates and selectivities in future generations of vitamin B6 enzyme models. An additional goal, which has not received adequate attention, is to generate high turnovers in reactions catalyzed by the vitamin B6 models.

Acknowledgments

We thank the students and postdocs of the Breslow laboratory who have worked over the years in this area, and whose names are in the references. The work has been supported by the NIH and NSF.

References

1 (a) R. A. John, *Biochim. Biophys. Acta* **1995**, *1248*, 81. (b) H. Hayashi, *J. Biochem.* **1995**, *118*, 463.

2 J. N. Jansonius, *Curr. Opin. Struct. Biol.* **1998**, *8*, 759.

3 (a) A. E. Martell, *Adv. Enzymol. Relat. Areas Mol. Biol.* **1982**, *53*, 163. (b) A. E. Martell, *Acc. Chem. Res.* **1989**, *22*, 115.

4 (a) A. E. Braunstein, M. M. Shemyakin, *Biohimiya (Moscow)* **1953**, *18*, 393. (b) D. E. Metzler, E. E. Snell, *J. Am. Chem. Soc.* **1952**, *74*, 919–983.

5 Some previous reviews: (a) R. Breslow, *Acc. Chem. Res.* **1995**, *28*, 146. (b) Y. Murakami, J.-i. Kikuchi, Y. Hisaeda, O. Hayashida, *Chem. Rev.* **1996**, *96*, 146. (c) R. Breslow, S. D. Dong, *Chem. Rev.* **1998**, *98*, 1997.

6 (a) T. C. Bruice, R. M. Topping, *J. Am. Chem. Soc.* **1962**, *84*, 2448. (b) T. C. Bruice, R. M. Topping, *J. Am. Chem. Soc.* **1963**, *85*, 1480. (c) T. C. Bruice, R. M. Topping, *J. Am. Chem. Soc.* **1963**, *85*, 1488.

7 S. C. Zimmerman, A. W. Czarnik, R. Breslow, *J. Am. Chem. Soc.* **1983**, *105*, 1694–1695.

8 R. Breslow, A. W. Czarnik, M. Lauer, R. Leppkes, J. Winkler, S. Zimmerman, *J. Am. Chem. Soc.* **1986**, *108*, 1969.

9 R. Breslow, J. Chemielewski, D. Foley, B. Johnson, N. Kumabe, M. Varney, R. Mehra, *Tetrahedron* **1988**, *44*, 5515.

10 J. Hine, *Acc. Chem. Res.* **1978**, *11*, 1.

11 S. C. Zimmerman, R. Breslow, *J. Am. Chem. Soc.* **1984**, *106*, 1490.

12 (a) K. Bernauer, R. Deschenaux, T. Taura, *Helv. Chim. Acta* **1983**, *66*, 2049. (b) R. Deschenaux, K. Bernauer, *Helv. Chim. Acta* **1984**, *67*, 373. (c) T. Chuard, F. Gretillat, K. Bernauer, *Chimia* **1993**, *47*, 215.

13 (a) H. Kuzuhara, T. Komatsu, S. Emoto, *Tetrahedron Lett.* **1978**, 3563. (b) T. Sakurai, H. Kuzuhara, S. Emoto, *Acta Crystallogr. B* **1979**, *35*, 2984. (c) M. Ando, Y. Tachibana, H. Kuzuhara, *Bull. Chem. Soc. Jpn.* **1982**, *55*, 829. (d) Y. Tachibana, M. Ando, H. Kuzuhara, *Chem. Lett.* **1982**, 1765. (e) Y. Tachibana, M. Ando, H. Kuzuhara, *Bull. Chem. Soc. Jpn.* **1983**, *56*, 3652. (f) Y. Tachibana, M. Ando, H. Kuzuhara, *Chem. Lett.* **1982**, 1769. (g) Y. Tachibana, M. Ando, H. Kuzuhara, *Bull. Chem. Soc. Jpn.* **1983**, *56*, 2263.

14 (a) M. Ando, H. Kuzuhara, *Bull. Chem. Soc. Jpn.* **1989**, *62*, 244. (b) M. Ando, H. Kuzuhara, *Bull. Chem. Soc. Jpn.* **1990**, *63*, 1925.

15 R. Breslow, M. Hammond, M. Lauer, *J. Am. Chem. Soc.* **1980**, *102*, 421.

16 R. Breslow, J. W. Canary, M. Varney, S. T. Waddell, D. Yang, *J. Am. Chem. Soc.* **1990**, *112*, 5212–5219.

17 R. Breslow, A. W. Czarnik, *J. Am. Chem. Soc.* **1983**, *105*, 1390.

18 J. Winkler, E. Coutouli-Argyropoulou, R. Leppkes, R. Breslow, *J. Am. Chem. Soc.* **1983**, *105*, 7198.

19 Y. Murakami, J. Kikuchi, O. Hayashida, *J. Inclusion Phenom.* **1989**, *7*, 91.

20 I. Tabushi, Y. Kuroda, M. Yamada, H. Higashimura, R. Breslow, *J. Am. Chem. Soc.* **1985**, *107*, 5545.

21 E. Fasella, S. D. Dong, R. Breslow, *Bioorg. Med. Chem.* **1999**, *7*, 709.

22 (a) H. Kondo, J. Kikuchi, J. Sunamoto, *Tetrahedron Lett.* **1983**, *24*, 2403. (b) H. Kondo, J. Kikuchi, S. Uchida, T. Kitamikado, E. Koyanagi, J. Sunamoto, *Bull. Chem. Soc. Jpn.* **1985**, *58*, 675. (c) J. Kikuchi, J. Sunamoto, H. Kondo, *J. Chem. Soc., Perkin Trans. 2* **1985**, 341. (d) H. Kondo, H. Tanamachi, J. Sunamoto, *Chem. Lett.* **1988**, 2013.

23 (a) Y. Murakami, J. Kikuchi, A. Nakano, K. Akiyoshi, T. Imori, *Bull. Chem. Soc. Jpn.* **1984**,

57, 1116. (b) Y. Murakami, J. Kikuchi, K. Akiyoshi, *Chem. Lett.* **1984**, 1185. (c) Y. Murakami, J. Kikuchi, K. Akiyoshi, *Bull. Chem. Soc. Jpn.* **1985**, *58*, 1200.

24 (a) Y. Murakami, J. Kikuchi, T. Imori, K. Akiyoshi, *J. Chem. Soc., Chem. Commun.* **1984**, 1434. (b) Y. Murakami, J. Kikuchi, K. Akiyoshi, T. Imori, *J. Chem. Soc., Perkin Trans. 2* **1986**, 1445.

25 (a) Y. Murakami, J. Kikuchi, K. Akiyoshi, T. Imori, *J. Chem. Soc., Perkin Trans. 2* **1985**, 1919. (b) Y. Murakami, J. Kikuchi, K. Akiyoshi, N. Shiratori, *Isr. J. Chem.* **1987/88**, *28*, 23. (c) Y. Murakami, J. Kikuchi, N. Shiratori, *J. Phys. Org. Chem.* **1989**, *2* 110.

26 (a) Y. Murakami, J. Kikuchi, N. Shiratori, *Bull. Chem. Soc. Jpn.* **1989**, *62*, 2045. (b) J. Kikuchi, Z.-Y. Zhang, Y. Murakami, *Chem. Lett.* **1994**, 1559. (c) J. Kikuchi, Z.-Y. Zhang, T. Miyajima, Y. Murakami, *Chem. Lett.* **1994**, 1701.

27 J. Kikuchi, Z.-Y. Zhang, Y. Murakami, *J. Am. Chem. Soc.* **1995**, *117*, 5383.

28 B. Imperiali, R. S. Roy, *J. Am. Chem. Soc.* **1994**, *116*, 12 083.

29 (a) B. Imperiali, R. S. Roy, *J. Org. Chem.* **1995**, *60*, 189. (b) R. S. Roy, B. Imperiali, *Tetrahedron Lett.* **1996**, *37*, 2129.

30 R. S. Roy, B. Imperiali, *Protein Eng.* **1997**, *10*, 691.

31 D. Qi, C.-M. Tann, D. Haring, M. D. Distefano, *Chem. Rev.* **2001**, *101*, 3081.

32 H. Kuang, M. L. Brown, R. R. Davies, E. C. Young, M. D. Distefano, *J. Am. Chem. Soc.* **1996**, *118*, 10 702.

33 H. Kuang, R. R. Davies, M. D. Distefano, *Bioorg. Med. Chem. Lett.* **1997**, *7*, 2055.

34 (a) H. Kuang, M. D. Distefano, *Protein Eng.* **1997**, *10*, 25. (b) H. Kuang, M. D. Distefano, *J. Am. Chem. Soc.* **1998**, *120*, 1072. (c) D. Qi, H. Kuang, M. D. Distefano, *Bioorg. Med. Chem. Lett.* **1998**, *8*, 875. (d) H. Kuang, D. Haring, D. Qi, A. Mazhary, M. D. Distefano, *Bioorg. Med. Chem. Lett.* **2000**, *10*, 2091. (e) D. Haring, M. D. Distefano, *Bioconj. Chem.* **2001**, *12*, 385. (f) D. Haring, M. D. Distefano, *Bioorg. Med. Chem.*

2001, *9*, 2461. (g) D. Haering, M. R. Lees, L. J. Banaszak, M. D. Distefano, *Protein Eng.* **2002**, *15*, 603.

35 L. Liu, R. Breslow, *J. Am. Chem. Soc.* **2002**, *124*, 4978.

36 L. Liu, M. Rozenman, R. Breslow, *J. Am. Chem. Soc.* **2002**, *124*, 12 660.

37 W. Zhou, L. Liu, R. Breslow, *Helv. Chim. Acta* **2003**, *86*, 3560.

38 L. Liu, R. Breslow, *J. Am. Chem. Soc.* **2003**, *125*, 12 110.

39 J. Chmielewski, R. Breslow, *Heterocycles* **1987**, *25*, 533.

40 L. Liu, R. Breslow, *Tetrahedron Lett.* **2001**, *42*, 2775.

41 J. Koh, PhD thesis, Department of Chemistry. Columbia University, 1994.

42 L. Liu, M. Rozenman, R. Breslow, *Bioorg. Med. Chem.* **2002**, *10*, 3973.

43 W. Weiner, J. Winkler, S. C. Zimmerman, A. W. Czarnik, R. Breslow, *J. Am. Chem. Soc.* **1985**, *107*, 4093.

44 (a) Y. Murakami, J. Kikuchi, T. Kitazaki, *J. Chem. Soc., Chem. Commun.* **1988**, 143. (b) Y. Murakami, J. Kikuchi, Y. Hisaeda, K. Nakamura, T. Kitazaki, H. Kaya, *Bull. Chem. Soc. Jpn.* **1990**, *63*, 2339.

45 (a) Y. Murakami, Y. Hisaeda, K. Nakamura, J. Kikuchi, *Chem. Lett.* **1990**, 1765. (b) Y. Murakami, Y. Hisaeda, T. Miyajima, H. Sakata, J. Kikuchi, *Chem. Lett.* **1993**, 645. (c) Y. Murakami, Y. Hisaeda, T. Miyajima, J. Kikuchi, *Chem. Lett.* **1993**, 2093.

46 (a) H. Kuzuhara, N. Watanabe, M. Ando, *J. Chem. Soc., Chem. Commun.* **1987**, 95. (b) M. Ando, J. Watanabe, H. Kuzuhara, *Bull. Chem. Soc. Jpn.* **1990**, *63*, 88.

47 J. T. Koh, L. Delaude, R. Breslow, *J. Am. Chem. Soc.* **1994**, *116*, 11 234.

48 (a) J. Kikuchi, T. Takashima, H. Nakao, K. Hie, H. Etoh, Y. Noguchi, K. Suehiro, Y. Murakami, *Chem. Lett.* **1993**, 553. (b) Y. Murakami, J. Kikuchi, T. Miyajima, Y. Hisaeda, *Chem. Lett.* **1994**, 55.

3
Evolution of Synthetic Polymers with Enzyme-like Catalytic Activities

Irving M. Klotz and Junghun Suh

3.1
Introduction: Conceptual Background

One of the most remarkable types of homogeneous catalyst is the class of naturally-occurring substances called *enzymes*. In the course of a billion years, nature has developed a range of macromolecules with high catalytic efficiency and extraordinary versatility in reactions whose rates can be accelerated. It has been a challenge for decades, therefore, to try to emulate the catalytic properties of enzymes with synthetic macromolecules of nonbiological origin (*synzymes*).

An enzyme is a macromolecule. It has long been recognized [1–3] that this macromolecule, M, operates through two sequential processes (Eq. 1): (1) binding of the substrate S and (2) provision of a molecular environment conducive to the chemical transformation:

$$M + S \underset{\text{step}}{\overset{\text{binding}}{\rightleftharpoons}} M \cdot S \xrightarrow[\text{step}]{\text{catalytic}} M + \text{products} \tag{1}$$

Particular enzymes show molecular selectivity in the binding step and provide specific functional groups and environments to facilitate the catalytic step.

In a rational approach to the reproduction of enzyme-like catalytic behavior with a synthetic polymer one should use, therefore, a two-step plan: (1) fabrication of binding sites on a suitable macromolecular framework; (2) introduction of groups to provide an environment of specific functional side chains that facilitate formation of the transition state in the catalytic step.

Artificial Enzymes. Edited by R. Breslow
Copyright © 2005 WILEY-VCH Verlag GmbH & Co. KGaA, Weinheim
ISBN: 3-527-31165-3

3.2
Homogeneous Polymer Biocatalysts

3.2.1
Fabrication of Macromolecules with Strong Affinities for Ligands

Endeavors toward this goal have been strongly guided by extensive earlier studies of binding of small molecules by proteins [4]. Serum albumin is extraordinary among non-enzymatic proteins in manifesting strong affinities for small molecules of widely different structure. Molecularly oriented and thermodynamic studies early disclosed the energetic quantities associated with its multiple, stepwise interactions. Associated free energies and entropies revealed the apolar and electrostatic interactions involved in the binding. Furthermore, changes in affinities in successive binding steps of a specific small molecule reflect macromolecular conformational adaptations.

Thus one might expect flexible, water-soluble synthetic polymers with suitable side chains to show affinities for small molecules. We have examined, over several decades, the binding ability of poly(vinylpyrrolidone), polyvinylpyridine, polylysine, polyacrylamide, poly(isopropylacrylamide), poly(vinylimidazole), poly(vinylmethyloxazolidinone), poly(vinylmethyloxazolidinone-vinylimidazole), poly(vinylpyrrolidone-vinylimidazole), poly(vinylpyrrolidone-vinyl alcohol), poly(vinylpyrrolidone-maleic anhydride), poly(vinylmethyloxazolidinone-maleic anhydride), and poly(2-methylaminoethyl methylacrylate-methacrylic acid). Other investigators have studied similar synthetic polymers [5–12]. In our experience no water-soluble polymer binds small molecules with an avidity comparable to serum albumin.

These polymers have high intrinsic viscosities (e.g. about 22 ml g^{-1} for polyvinylpyrrolidone), which indicate that the macromolecules are swollen and extended in water. In contrast, serum albumin, with an intrinsic viscosity near 4 ml g^{-1}, must be relatively compact. Such a relatively compact conformation might be obtained with a water-soluble polymer by introducing cross-linkages or by using a highly branched matrix. Indeed, the latter has proved to be particularly fruitful.

An interesting polymer constrained to a relatively compact conformation is polyethylenimine (PEI), which, as usually prepared, gives a highly branched rather than a linear macromolecule. Figure 3.1(A) shows the structure of a segment of this polymer. Approximately 25 % of its nitrogens are primary amines, 50 % secondary, and 25 % tertiary [13]. The polymer branching may be represented schematically (Figure 3.1B).

The primary amine groups form very suitable loci for the attachment of apolar groups. We have, therefore, prepared several derivatives with different side chains attached to a portion of the primary amine groups. These modified polymers show remarkable binding properties [14]. Figure 3.2 illustrates the tremendously greater extent of binding by the acylpolyethylenimines as compared with serum albumin. At a free ligand concentration of 10^{-5} M, the lauroyl derivative of polyethylenimine binds over 100 moles of small dye molecule, the hexanoyl derivative about 10, and the butyryl about 1 (Figure 3.2), whereas studies with bovine albumin lead to values just below 1.

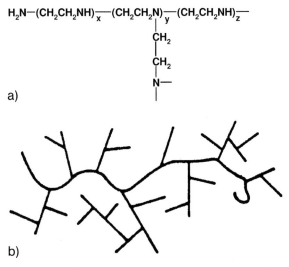

a)

b)

Figure 3.1 (a) Structure of a segment of polyethylenimine (PEI). (b) Schematic representation of multiple branching in PEI.

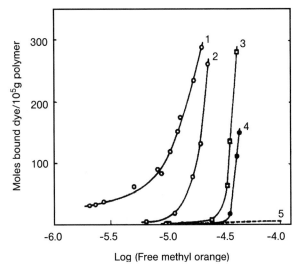

Figure 3.2 Extent of binding of methyl orange at pH 7.0 and 25 °C as a function of free (nonbound) dye concentration. (1) Polyethylenimine (PEI) with 8.4 % of residues acylated by lauroyl groups. (2) PEI with 11.5 % of residues acylated by hexanoyl groups. (3) PEI with 10 % of residues acylated by butyryl or isobutyryl groups. (4) PEI-600. (5) Bovine serum albumin.

3.2.2
Enhanced Reactivity of Nucleophiles in Polyethylenimines (PEIs)

The numerous amine nitrogens of polyethylenimine can act as nucleophiles. In addition, they are locally concentrated (Figure 3.1). Furthermore, polymers with attached aliphatic acyl groups provide apolar binding sites in proximity to the amine residues. One might expect, therefore, to find progressively enhanced rates of aminolysis of substrates with increasingly large apolar substituents. Quantitative measurements of aminolysis rates indeed revealed enhancements of several orders of magnitude.

To assess the aminolytic effectiveness of the polymer amines, the cleavage reaction of Eq. (2) has been followed. If an amine $P-NH_2$ is used in the aqueous solution, one obtains RCONHP instead of RCOOH.

$$O_2N-\!\!\left\langle\bigcirc\right\rangle\!\!-O-\overset{\overset{\displaystyle O}{\|}}{C}\!\!-R \quad\xrightarrow{\ H_2O\ }\quad O_2N-\!\!\left\langle\bigcirc\right\rangle\!\!-OH \ + \ HO-\overset{\overset{\displaystyle O}{\|}}{C}\!\!-R \tag{2}$$

Table 3.1 [15] lists the first-order rate constants, which are corrected for hydrolysis of ester in buffer alone. Propylamine served as a reference amine; in its presence $10^2 k$ (min^{-1}) for aminolysis decreased progressively from 0.98 to 0.51 to 0.05 as the acryl group increased from 2 to 12 carbons (Table 3.1). The sharp drop for nitrophenyl laurate may be the result of micelle formation (even at 6×10^{-6} M).

With nonacylated polyethylenimines (Table 3.1) the rate constant is increased ca. 4-fold over that of propylamine. This small enhancement may merely be because a greater fraction of the polymer's primary amine groups are in the basic, NH_2 state. With these polyethylenimines, as with propylamine, k drops with increasing length of the hydrocarbon chain of the acyl nitrophenyl ester.

Markedly different trends in the rate constants appear for aminolysis by lauroyl-polyethylenimine (containing 10 residue % lauroyl groups). For each nitrophenyl ester the rate is substantially greater with lauroyl-polyethylenimine than with polymer contain-

Table 3.1 First-order rate constants $(k \times 10^2\ min)^a$ for amine acylations by p-nitrophenyl esters.[b]

Amine[c]	p-Nitrophenyl acetate	p-Nitrophenyl caproate	p-Nitrophenyl laurate
Propyl	0.98	0.51	0.053
PEI-6	3.60	1.47	0.11
PEI-18	4.38	1.57	0.11
PEI-600	4.60	1.80	0.17
L(10%)-PEI-6[d]	15.2	68.1	698

a) Here $k = k_a - k_o$, where k_a is the measured rate constant in the presence of amine and k_o is that for the hydrolysis in tris buffer alone; k_o is 0.94 $\times 10^{-2}\ min^{-1}$ for the acetyl ester, 0.61 $\times 10^{-2}\ min^{-1}$ for the caproyl ester, and 0.023 $\times 10^{-2}\ min^{-1}$ for the lauroyl ester. Measurements made at pH 9.0 in 0.02 M tris(hydroxymethyl)-aminomethane buffer, 25 °C. Stock solutions of substrate were made in acetonitrile; hence the final buffer also contained 6.7% acetonitrile.
c) Numeral following polyethylenimine multiplied by 100 is the molecular weight of the polymer sample.
d) This sample of PEI-6 has 10% of its nitrogens acylated with lauroyl groups.

ing no acyl group. Furthermore, k now increases markedly as the acyl group is lengthened from 2 to 12 carbons (Table 3.1). For nitrophenyl laurate, k with lauroyl-polyethylenimine is 10^4 times greater than that with propylamine. Such a comparison may not be fully appropriate if the low rate with the reference amine is due primarily to the micellar state of the lauroyl nitrophenyl ester. If one assumes that in the absence of micelle formation the long-chain ester would show a rate comparable to that of acetyl nitrophenyl ester, the enhancement factor in the presence of lauroyl-polyethylenimine is still of the order of 10^3. In any event, the introduction of strong binding sites on the polymer clearly leads to marked rate enhancements.

3.2.3
Polyethylenimines with Nucleophile Adducts

Having polymers with strong ligand binding abilities, and with capacities to accelerate rates, one can graft onto the macromolecular matrix truly catalytic functional groups. The imidazole moiety is an attractive, well-recognized nucleophile (particularly effective in model systems) in catalyzing the hydrolysis of nitrophenyl esters (Eq. 2).

With a polyethylenimine containing 10% of its residues alkylated with dodecyl groups and 15% alkylated with methyleneimidazole substituents, esterolysis is truly catalytic [16]. Table 3.2 compares the catalytic effectiveness of this polymer biocatalyst (synzyme) with that reported for other substances that accentuate nitrophenyl ester hydrolysis [17, 18]. Clearly, this polymer is nearly 300 times as effective as free imidazole, but it does not match chymotrypsin, even with the activated unnatural nitrophenyl ester substrate, let alone peptide substrates.

Nucleophiles other than imidazole have also been coupled to polyethylenimines and other polymers. These include hydroxamate [19–23], mercaptan [24, 25], aldoxime [26], and triazine [27] functionalities. In general, these moieties are active in transferring the acyl group of nitrophenyl esters to the nucleophile, but deacylation of the acylnucleophile is slow.

Conversely, dimethylaminopyridines attached to polymers are remarkably effective catalysts of the solvolysis of esters, with true turnover [28–30].

Pyridine itself attached to polyethylenimines (through a reduced Schiff base linkage) shows no catalysis, evidently because in the polymer environment both the pK_a and nucleophilicity of the ring nitrogen are reduced substantially. However, 4-N,N-dialky-

Table 3.2 Relative effectiveness of various catalysts in cleavage of nitrophenyl esters.

Catalyst	Catalytic Constant k (M^{-1} min^{-1})
Imidazole	10^a
α-Chymotrypsin	$10\ 000^b$
PEI-600, 10% dodecyl. 15% imidazole groups	2700

a) Substrate was p-nitrophenyl acetate, at a pH near neutrality. Taken from Ref. 17.
b) Taken from Ref. 18.

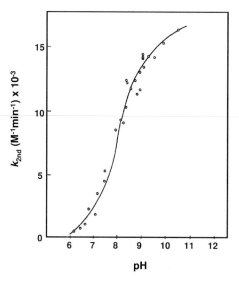

1

laminopyridines are far superior to pyridine as acylation catalysts [31, 32], and were promising candidates for attachment to modified polyethylenimines with hydrophobic substituents. Various substituted aminopyridines of type **1** have been coupled to poly-ethylenimines and the polymers obtained are very effective catalysts of the solvolysis of nitrophenyl esters [28, 30].

The rate–pH curve for these polymers is S-shaped through the accessible range (Figure 3.3). The inflection point occurs at about pH 8.0 due to an average pK_a of the polymer-bound pyridine, but the small molecule dialkylaminopyridine shows a pK_a of 9.7. The cationic charge and apolar environment of the polymer favor the non-charged form of the heterocycle adduct. Since binding of a non-charged substrate by the polymer depends on apolar interactions, not electrostatic ones, the loss in charge on the polymer with increasing pH does not diminish binding and, hence, no bell-shaped curve appears.

The pK_a of the aminopyridines attached to the polymer (Table 3.3) are invariably lower than those of the corresponding isolated small molecules. The cationic character of polyethylenimine produces an electrostatic field that weakens the basicity of the nucleophile by 2–3 pK units. A lowering of an additional unit ocurrs when a long apolar group is attached to the aminopyridine moiety. Evidently, in this molecule the apolar environment favors further the uncharged, nonprotonated form of the pyridine.

Figure 3.3 Rate (second order)–pH profile for hydrolysis of p-nitrophenyl caproate in aqueous solution at 25 °C by $(C_2H_4)_m(C_{12}H_{25})_{0.12m}(C_9H_{11}N_2O)_{0.10m}$, a 60 000 molecular weight polyethylenimine with 10% of the nitrogens covalently attached (by an amide bond) to 3-[N-methyl-N-(4-pyridyl)amino]propionic acid and 12% linked to lauryl groups.

Table 3.3 Properties of polyethylenimines with covalently-linked aminopyridines.

Polymer derivative[a]	pK_a	Second-order rate constants[b]			
		pH 7.3		pH 9.2	
		k_2(obs.)	k_2(norm.)	k_2(obs.)	k_2(norm.)
PEI-Pyr-1	7.31	10 400	20 800	34 300	36 400
PEI-Pyr-2	6.64	14 700	25 000	53 300	55 500
PEI-Pyr-3	6.68	25 300	35 200	164 000	168 000

a) Stoichiometric composition given: C_2H_4N represents ethylenimine residue and $(C_{12}H_{25})$ lauryl group. For PEI-600, m approximately 1400.

b) Observed rate, (obs.); rate normalized to non-protonated pyridine (norm.) adduct.

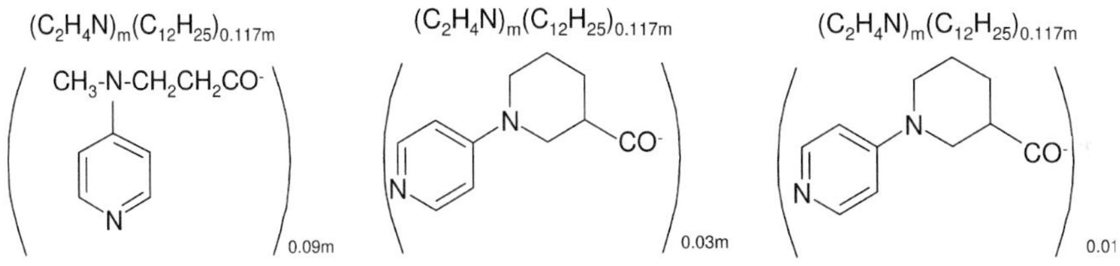

PEI-Pyr-1 **PEI-Pyr-2** **PEI-Pyr-3**

For these adducts, the polymer environment increases markedly the concentration of the nucleophilic state of the pyridine nitrogen. Unsurprisingly, therefore, in every case [30] the observed hydrolysis rates of *p*-nitrophenyl caproate (pH 7.3) are 50–2000-fold greater than that for the isolated aminopyridine. The nucleophilic forms of the small molecule aminopyridines are about 200 times more effective than imidazole.

The catalytic effectiveness of the small molecule or of the polymer adducts is increased when the pH of the solution is raised from 7.3 to 9.2.

The influence of the polymer on the *intrinsic* effectiveness of the nonprotonated pyridine moiety adduct can be assessed by comparison of the normalized rates k_2(norm.), i.e., those expressed in terms of the concentration of $>N-$ species in the solution. The intrinsic activity of the most effective nucleophiles (Table 3.3) is enhanced by a factor of 35, when it is attached to polyethylenimine. This increase is in addition to that due to the ability of the polymer at any pH to increase the fraction of aminopyridine in the uncharged $>N-$ form. For a specific nucleophile, for which polymer adducts have been prepared with different aminopyridine contents, the catalytic effect, per pyridine, increases with decreasing residue concentration of the nucleophile (Table 3.3).

The acceleration in cleavage rate of nitrophenyl caproate by the dialkylaminopyridine polymers was shown to be a manifestation of true catalysis by complete hydrolysis of a 50-fold excess of substrate over residue molar concentrations of pyridine moiety attached to the polymer.

3.2.4
Proximal Group Adducts to Polyethylenimines

Peptide bonds are very stable, having a half-life for spontaneous hydrolysis at pH 7 and 25 °C of 500–1000 years [33, 34]. Many proteases hydrolyze peptide bonds by using several organic functional groups such as carboxyl, imidazolyl, hydroxyl, and mercapto groups as catalytic elements. Examples are aspartic proteases, such as pepsin and HIV protease, for which the active site contains two aspartic carboxyl groups as the key catalytic entities for peptide hydrolysis. To synthesize an enzyme-like polymer with high catalytic activity [35] one needs to design an active site composed of several catalytic groups in close proximity.

Since the 1960s, numerous efforts have been made to design organic catalysts for the hydrolysis of peptides as models of proteases. In most cases, one could only succeed in hydrolyzing activated analogues of peptides. The first synthetic organic compound with peptidase activity was obtained by constructing an active site consisting of three convergent salicylate residues attached to the backbone of polyethylenimine [36]. As illustrated in Eq. 3, three molecules of 4-bromoacetylsalicylate were assembled around a Fe(III) ion and the module was then cross-linked with polyethylenimine (PEI). Once an amino group of PEI was linked to one of the three salicylates through a covalent bond, neighboring PEI amino groups, located in the right positions, reacted with the remaining two salicylates, thereby completing the cross-linkage. After removal of Fe(III) by treatment with acid, a water-soluble polymer with sites (Sal₃), consisting of three proximal salicylates, was obtained.

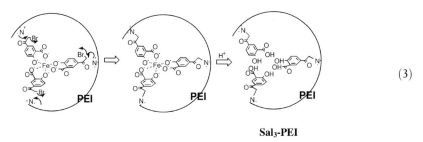

$$(3)$$

Sal₃-PEI

0   1   2   3   4   5   6   10   24   72   hr

Figure 3.4 Electrophoretic analysis of γ-globulin (5.35×10^{-6} M) incubated with (Sal)₃-PEI (total concentration of (Sal)₃ sites: 4.95×10^{-4} M) at pH 6.00 and 50 °C. Numbers indicate the period of incubation. Upper and the lower bands are the heavy and the light chains of γ-globulin.

Figure 3.5 Dependence on pH of the second-order rate constant (k_2) for the hydrolysis of the heavy (●; curve b) and the light (□; curve a) chains of γ-globulin catalyzed by (Sal)$_3$-PEI at 50 °C.

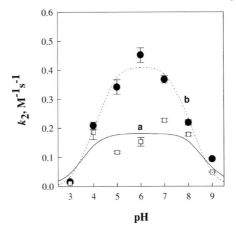

The Sal$_3$ site contains three carboxyl and three phenol groups in addition to amino groups provided by the polyethylenimine backbone. Since such functionalities are active groups in enzymes, they might provide a catalytic module in Sal$_3$-PEI. When Sal$_3$-PEI was incubated with γ-globulin, protein degradation was observed by SDS-PAGE electrophoresis (Figure 3.4). Kinetic data, collected by measuring the decrease in intensity of the electrophoretic band, revealed a half-life as short as 1 h at pH 7 and 50 °C for this γ-globulin degradation. When salicylates were attached to polyethylenimine randomly, little catalytic activity was observed, indicating that the proteolytic activity of Sal$_3$-PEI arose from collaboration among proximal salicylates. The pH dependence of the second-order rate constant (Figure 3.5) revealed optimum activity at pH 6–7. Since carboxyl, phenol, and amino groups can all play catalytic roles at pH 6–7, one cannot tell which of these functional groups is responsible for the proteolysis.

In the preparation of Sal$_3$-PEI, each building block of the preassembled module is linked to polyethylenimine by a single attachment. This would allow flexibility of the catalytic moieties incorporated into the active site. Conversely, double attachment of the building blocks preassembled by the scaffold to the polyethylenimine framework may better conserve the original geometry of the preassembled module.

Highly effective host molecules for metal ions have been designed with 2,9-bis(bromomethyl)-1,10-phenanthroline (**2**) by the cross-linkage method [37]. By the reaction

2

of polyethylenimine with Cu(I)(2)$_2$, two molecules of **2** preassembled by Cu(I) ion were cross-linked with polyethylenimine to obtain [Cu^{I}Phen$_2$]^{PA}PEI (Eq. 4). By the reaction of **2** with polyethylenimine in the absence of Cu(I) ion, phenanthroline moieties were attached randomly to polyethylenimine to obtain [Phen]^{Ran}PEI, and then the primary and secondary amino groups of polyethylenimine were acetylated to obtain [Phen]$^{Ra-n}$AcPEI.

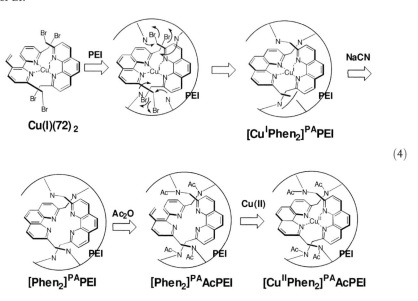

Cu(I)(72)$_2$ [Cu^{I}Phen$_2$]^{PA}PEI

(4)

[Phen$_2$]^{PA}PEI   [Phen$_2$]^{PA}AcPEI   [Cu^{II}Phen$_2$]^{PA}AcPEI

The two preassembled phenanthrolines cross-linked by PEI would remain in close proximity when the conformational freedom of the resulting two macrocycles is sufficiently suppressed. If the two phenanthroline moieties are very close, they will affect the protonation of each other considerably. For [Phen]^{Ran}PEI and [Phen]^{Ran}AcPEI, the phenanthroline moieties attached randomly to the polymer behaved as monobasic species (Eq. 5) with a pK_a of 1.2. For [Phen$_2$]^{PA}PEI or [Phen$_2$]^{PA}AcPEI, the pair of phenanthrolines originally preassembled by Cu(I) behaved as one ionizing unit (Eq. 6) with two pK_as of ca. 1.3 and ca. 10.8. The pK_{a2} of [Phen$_2$]^{PA}PEI or [Phen$_2$]^{PA}AcPEI is greater by ca. 9.5 pK_a units than the corresponding pK_{a1}. This pK_a difference is attributable to destabilization of the diprotonated form of Eq. (6) by electrostatic interaction between the two cationic sites and stabilization of the corresponding monoprotonated form by hydrogen bonding between two phenanthroline moieties. Both these interactions originate from the close proximity of the two preassembled phenanthrolines.

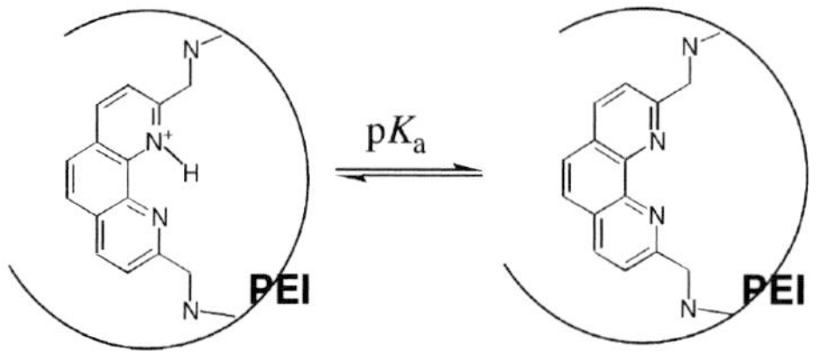

(5)

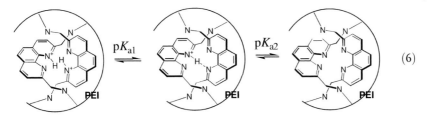

(6)

Close proximity between the phenanthrolines in $[Phen_2]^{PA}AcPEI$ would lead to strong binding of metal ions. For the Cu(II) complex of $[Phen_2]^{PA}AcPEI$, log K_f was 17.0 when the phenanthrolines were fully deprotonated. The effectiveness of cooperation between two phenanthroline moieties within a binding site for the Cu(II) ion of $[Phen_2]$-$^{PA}AcPEI$ may be expressed in terms of effective molarity, an idea originally introduced as a measure of the efficiency of intramolecular compared with intermolecular catalysis [38]. The effective molarity of a phenanthroline toward a Cu(II) ion bound to the other phenanthroline in the phenanthroline pair (**3**) was estimated as 10^6 M.

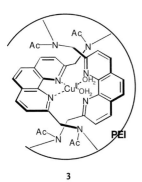

3

The effective molarity measured for $[Phen_2]^{PA}AcPEI$ may be compared with that for enterobactin, the strongest microbial siderophore, containing three catechol units connected by a spacer. The effective molarity of a catechol unit towards a Fe(III) ion bound to another catechol unit contained in enterobactin is estimated to be 3×10^4 M [37]. Enterobactin contains three catechol units, whereas the Cu(II) binding site of $[Phen_2]$-$^{PA}AcPEI$ consists of only two phenanthrolines. Nevertheless, the effective molarity observed for $[Phen_2]^{PA}AcPEI$ is extraordinary for a synthetic system.

Designing active sites containing functional moieties with high effective molarities is necessary for the construction of effective synthetic molecules with high enzyme-like or antibody-like activities. Cross-linkage of building blocks preassembled by a scaffold by single or double attachment to PEI produced synthetic polymers effective for catalysis or molecular recognition.

3.2.5
Polyethylenimines (PEIs) with Adducts that Self-assemble into Catalytic Moieties

Metal ions participate in the catalytic action of many enzymes. In the hydrolysis of peptide bonds of proteins and phosphodiester bonds of RNA and DNA, for example, metal ions play various catalytic roles, acting as Lewis acid catalysts [39]. 2-Hydroxy-propyl-*p*-nitrophenylphosphate (HPNPP) has been widely employed as an activated analogue of RNA. Transesterification of HPNPP is effectively catalyzed by two or more metal ions acting together [40, 41]. This catalytic action has been explained in terms of activation of the two phosphoryl–oxygen bonds and general base assistance for the intramolecular attack of the hydroxyl group (**4**) [41].

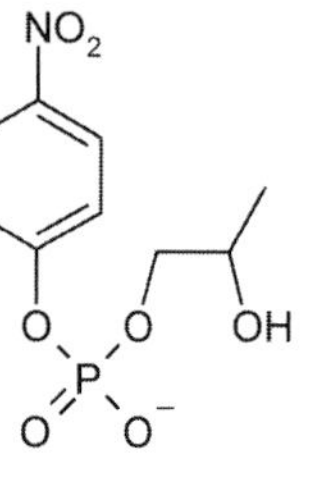

HPNPP 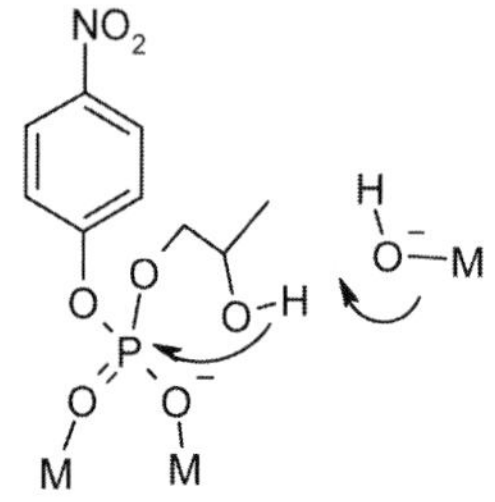   **4**

As rationally designed multinuclear catalysts for transesterification of HPNPP, metal complexes such as **5** have been designed [42]. By random attachment of lauryl groups and Ni(II) complexes of terpyridine (TP) to PEI (**6**) we have synthesized a polymer that has much higher catalytic activity than **5** for the same reaction [43]. The contents of the pendant groups were varied in several combinations, and the best catalyst ([Ni(II)TP]$_5$Lau$_{12}$PEI) contained 5 residue mol% Ni(II)TP and 12 residue mol% lauryl group.

The calix[4]arene-based binuclear Zn(II) complex (**5**) manifests optimum activity at pH 7.4. The maximum k_o attainable by **5** is 1.0×10^{-3} s^{-1} regardless of pH or C_o. Much higher k_os were observed with [Ni(II)TP]$_5$Lau$_{12}$PEI (Figure 3.6).

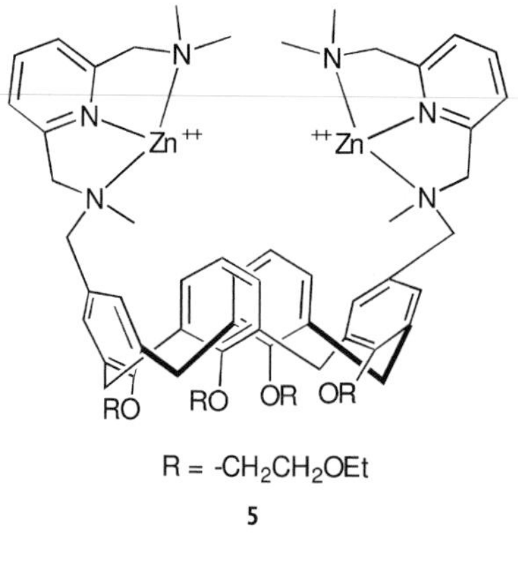

R = -CH$_2$CH$_2$OEt

5

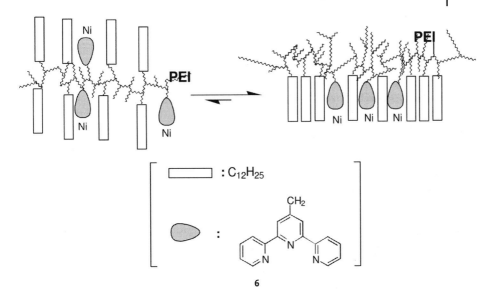

6

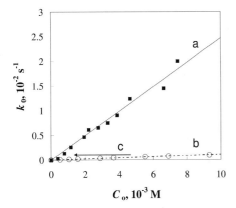

C_o, 10^{-3} M

Figure 3.6 k_o vs. C_o (total concentration of Ni^{II}TP) for [Ni^{II}TP]$_5$Lau$_{12}$PEI at 25 °C and pH 9.50 (a; ■) or 7.40 (b; ○). The arrow (c) represents the maximum rate constant attainable with **5** regardless of pH or C_o.

7

In water, the lauryl and terpyridyl residues are expected to form hydrophobic clusters (**6**) on the backbone of PEI. An active site consisting of two or more metal ions can form in the cluster. Its structure depends on the contents of lauryl and terpyridyl residues. An active site obtained by combinatorial attachment of the pendants catalyzed transesterification of HPNPP apparently by using adjacent metal centers (**7**). In this work, we created the active site by self-assembly of hydrophobic pendants attached to the backbone of PEI, taking advantage of the branched and flexible structure of the polymer.

3.3
Heterogeneous Polymer Biocatalysts

3.3.1
Random Catalytic Adducts

Branched insoluble polymers have more rigid backbones and more hydrophobic microdomains than branched soluble polymers such as PEI, which can be advantageous in designing some types of catalysts. In addition, such insoluble catalysts may be more useful in practical applications.

An effective polymer catalyst with protease activity has been obtained simply by attaching imidazoles randomly to poly(chloromethylstyrene-*co*-divinylbenzene) (PCD). This polymer is a cross-linked polystyrene (PS) with the styryl moiety being chloromethylated. Imidazole was attached to PCD through either C-2 or the N atom [44]. Unreacted chloro groups of PCD were blocked by treatment with methoxide. In the best catalyst we obtained ($[Im^C]_{22}$PCD) imidazole was attached to PCD via C-2 and the content of imidazole was 22 residue mol%.

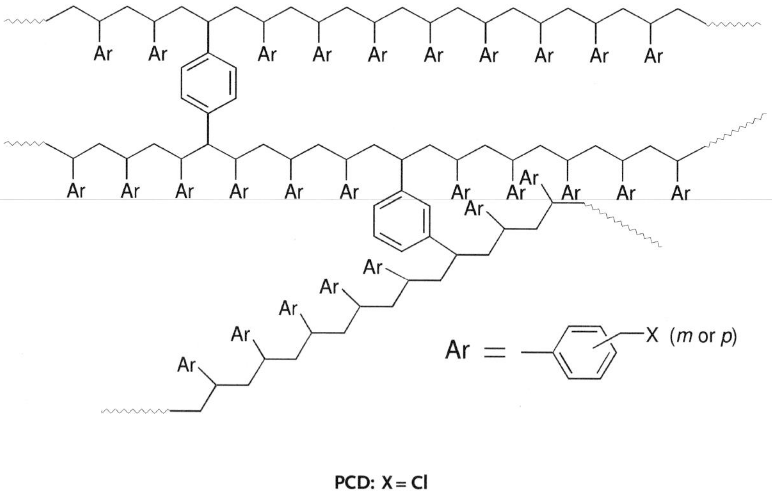

PCD: X = Cl
PAD: X = NH$_2$

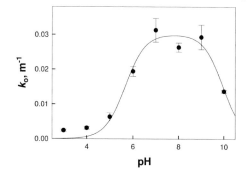

Figure 3.7 pH profile of k_o for the hydrolysis of albumin ($S_o = 1.50 \times 10^{-6}$ M) by $[Im^C]_{22}PCD^{MeO}$ at 25 °C ($C_o = 0.115$ M: hereafter, C_o for a heterogeneous catalyst represents the total concentration of the catalytic center or the catalytic module attainable when the catalyst is assumed to be dissolved).

Polymer $[Im^C]_{22}PCD$ was effective in hydrolyzing albumin, manifesting optimum activity at pH 7–9 (Figure 3.7). The highest k_o shown in Figure 3.7 corresponds to a half-life of 20 min at pH 7 and 25 °C. Proteolytic activity was reduced markedly (Figure 3.8) when the imidazole content attached via C-2 is lowered from 22 to 5 residue mol% or when imidazole was attached via the N atom. This indicates that the active site on $[Im^C]_{22}PCD^{MeO}$ contains two or more imidazole moieties. Eq. (7) presents a mechanism consistent with the catalytic action of two imidazoles with unalkylated N atoms and with the pH profile of Figure 3.7.

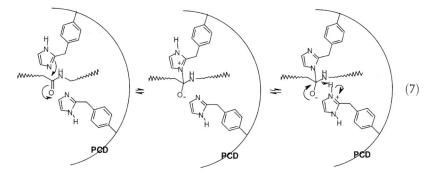

$$(7)$$

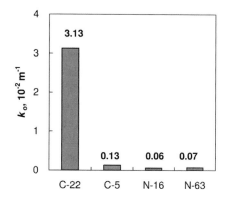

Figure 3.8 Comparison of k_o for the hydrolysis of albumin ($S_o = 1.50 \times 10^{-6}$ M) by $[Im^C]_{22}PCD^{MeO}$ (C-22), $[Im^C]_5PCD^{MeO}$ (C-5), $[Im^N]_{16}PCD^{MeO}$ (N-16), and $[Im^N]_{63}PCD^{MeO}$ (N-63) ($C_o = 0.115$ M) at pH 7.00 and 25 °C. In the nomenclature of the catalysts, superscript C or N indicates that imidazole was attached via the C or N atom, respectively, and the subscript is the residue mol% content of imidazole attached to PCD.

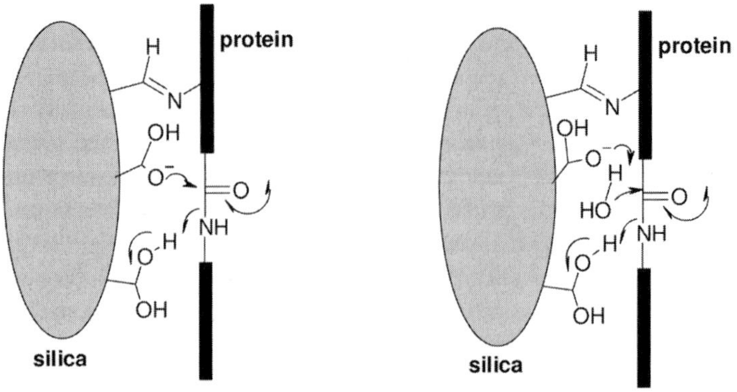

8

Carboxyl and imidazolyl groups are the catalytic moieties present in the side chains of amino acids. However, protease-like catalysts could be designed with organic functional groups unrelated to amino acids. Indeed, a protease-like catalyst was synthesized by using aldehyde groups [45]. Silica gel covered densely with aldehyde groups manifested catalytic activity toward albumin and γ-globulin with half-lives as short as 1.5–3 h at pH 5–6 and 50 °C. As proposed in **8**, aldehydes can act as binding sites by forming imines with amino groups located on the protein surface. In addition, the hydrates of aldehydes can act as general acids, general bases, or nucleophiles to hydrolyze the peptide bonds of the bound protein substrate. Catalytic action of an aldehyde hydrate in a transacylation reaction has been reported [46].

When the surface of silica gel was covered with organic functional groups such as carboxyl, hydroxyl, phenolic, imidazolyl, or mercapto groups as well as aldehyde groups, the proteolytic activity was considerably improved [47]. Silica gel covered with histidines and aldehydes afforded the best activity: the half-life was as short as 50 min at pH 6–9 and 25 °C. Furthermore, broad substrate specificity was achieved with various proteins (albumin, ovalbumin, hemoglobin, and γ-globulin) being hydrolyzed. Here, the aldehyde groups would act as the binding site and imidazolyl, carboxyl, and/or aldehyde hydrates would provide catalytic groups in a mechanism analogous to **8**.

3.3.2
Proximal Group Adducts

Instead of random attachment of functional groups to the solid supports, active sites consisting of two or more organic catalytic groups have been created on polystyrene backbones. For example, poly(aminomethylstyrene-*co*-divinylbenzene) (PAD), the cross-linked polystyrene with styryl residues containing aminomethyl groups, provides a framework for attachment of an active site that consists of three salicylate groups. A procedure similar to that shown in Eq. (3) has been used to build such a heterogeneous catalyst [48]. The excess amino groups on the resulting polystyrene were acetylated and then Fe(III) ion was removed with acid to obtain Sal$_3$-PAD-Ac.

Albumin was effectively hydrolyzed by Sal$_3$-PAD-Ac, with optimum activity at pH 3 – just as in the action of pepsin, the most typical aspartic protease. Protease action of Sal$_3$-PAD-Ac at pH 3 appears to involve collaboration of proximal carboxyl groups, in analogy with pepsin.

Active sites with peptidase activity were also created on polystyrene backbones with attached metal complexes as the catalytic center. When the Cu(II) complex of cyclen (Cyc) was attached to cross-linked polystyrene, the proteolytic activity of Cu(II)Cyc was enhanced remarkably [49]. By substitution of the chloro groups of PCD with various nucleophiles, PCD derivatives **9** and **10** were prepared. γ-Globulin was hydrolyzed effectively upon incubation with **9** and **10**. Proteolytic activity of Cu(II)Cyc was enhanced by up to 10^4-times upon attachment to the polystyrene. Since only a small fraction of Cu(II)Cyc moieties is present on the open surface on PCD and can participate in the hydrolysis of γ-globulin, the normalized degree of activation must be substantially greater than 10^4-fold. Activation of Cu(II)Cyc on the surface of polystyrene was attributed to the hydrophobic environment.

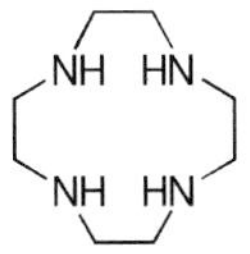

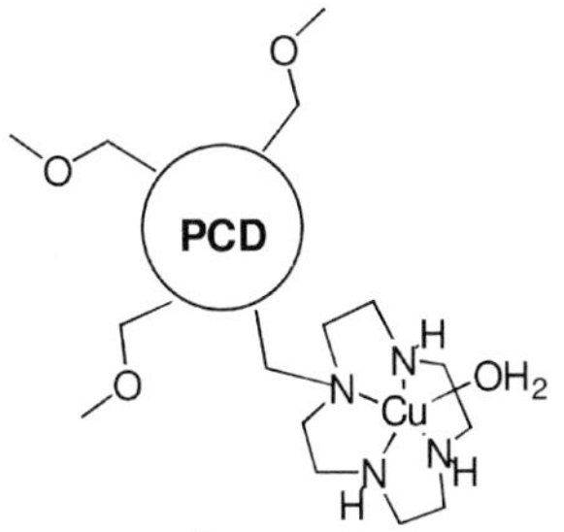

Since reactive metal centers were secured, attempts were initiated to achieve substrate selectivity in amide hydrolysis by the metal complexes [50, 51]. The active site of **11** was constructed on the surface of partially chloromethylated cross-linked polystyrene (PCPS) by stepwise modification of the polymer [50]. Here, the active site was chiral since L-Arg was used to link the guanidinium group to the polymer. Several cinnamoyl amide derivatives (**12–15**) were tested as substrates for **11**. Neutral amide **12** was not hydrolyzed upon incubation with **11**, but carboxyl-containing amides **13–15** were hydrolyzed effectively by **11**, with the optimum activity at pH 9 (Figure 3.9). Also hydrolyzed by **11** were both acetyl L-Phe and acetyl D-Phe; a small (1.5) enantioselectivity was observed as the catalyst is chiral.

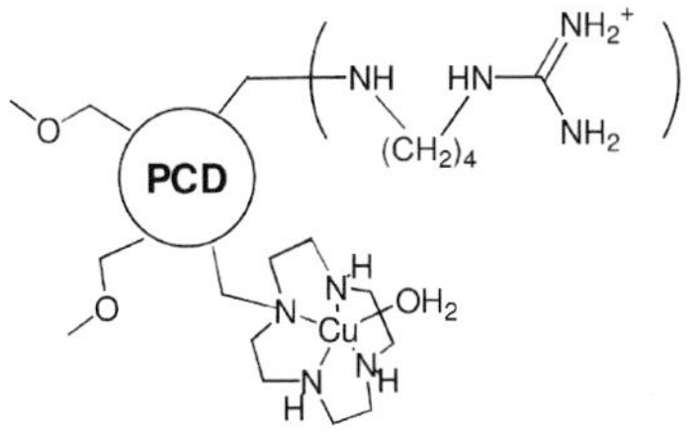

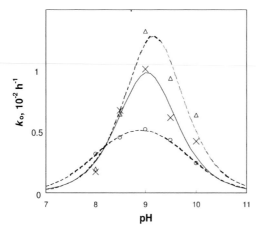

12: R = CH$_3$
13: R = CH$_2$COOH
14: R = CH$_2$CH$_2$COOH
15: R = CH$_2$CH$_2$CH$_2$COOH

In the proposed mechanism of hydrolysis (**16**), the carboxylate anion of **13–15** or acetyl Phe is recognized by the guanidinium ion, and the Cu(II) center subsequently hydrolyzes the amide group. Both the electrostatic interaction between carboxylate and guanidinium ions and the Cu(II)-catalyzed amide cleavage would be facilitated by the microenvironment provided by polystyrene.

In carboxypeptidase A [52, 53], the active-site Zn(II) ion plays essential catalytic roles and the guanidinium of Arg-145 recognizes the carboxylate anion of the substrates, thus making the enzyme an exopeptidase. Important features of carboxypeptidase A reproduced by **11** include the essential catalytic action of a metal ion and participation of a guanidinium group in substrate recognition, so that this polymer biocatalyst hydrolyzes unactivated amides, and exhibits selectivity toward amide bonds adjacent to carboxylate groups in the substrate.

Multinuclear metal centers are present in the active sites of several metalloproteases. A catalytic polymer with peptidase activity containing trinuclear active sites has been prepared [51] using **17**. Upon treatment of **17** with excess NaH, at least three of the six

Figure 3.9 pH dependence of k_o for hydrolysis of **13** ($\triangle$), **14** ($\times$), and **15** ($\bigcirc$) ($S_o = 1.96 \times 10^{-4}$ M) promoted by **11** ($C_o = 1.08$ mM) at 50 °C. Unlike the carboxyl-containing amides, the neutral amide (**12**) was not hydrolyzed.

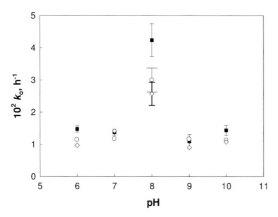

16

N–H protons were deprotonated. The anion of **17** was attached to PCD by mixing it with the polymer; **17** is thus triply attached to PCD (**18**). After the chloro groups of **18** were displaced with methoxide, the resulting resin was treated with acid and then with CuCl$_2$ to produce **19**. The active site in **19** contains three convergent Cu(II) complexes of tris(2-aminoethyl)amine. On incubation with **19**, neutral amide substrate **12** was unaffected whereas carboxyl-containing amides **13–15** were hydrolyzed effectively (Figure 3.10). The mechanism depicted in **20** was proposed to account for the substrate selectivity manifested for the carboxyl-containing amides.

The k_{cat} estimated for the amide hydrolysis by **11** or **19** was much greater than either 0.1 h^{-1} at pH 8.5 and 50 °C or 0.2 h^{-1} at pH 8 and 50 °C. This k_{cat} may be compared with the 0.18 h^{-1} for hydrolysis of an amide substrate (at the optimum pH of 9 and 25 °C) observed with a catalytic antibody [54] that had been elicited by a joint hybridoma and combinatorial antibody library approach.

Figure 3.10 pH dependence of k_o for the hydrolysis of **13–15** ($S_o = 1.96 \times 10^{-4}$ M: ■, **13**; ○, **14**; ◇, **15**) catalyzed by **19** ($C_o = 5.80 \times 10^{-4}$ M) at 50 °C. Unlike the carboxyl-containing amides, the neutral amide (**12**) was not hydrolyzed.

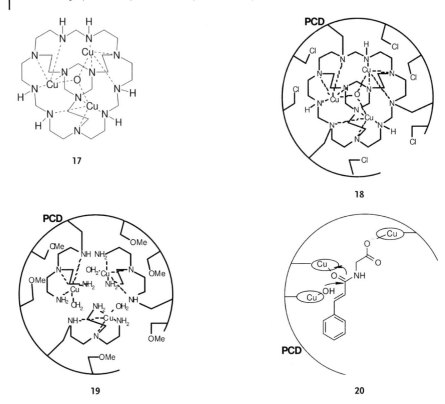

3.3.3
Adducts Containing Catalytic Modules Synthesized Prior to Incorporation into Polymers

To develop a methodology applicable to the design of a wide range of multinuclear active sites on the backbones of insoluble polymers we prepared a molecular entity, composed of various catalytic elements, with a precisely defined structure and then attached it to a polymeric backbone. Thus, we synthesized catalytic modules containing one, two, or four metal-chelating sites, which were subsequently attached to a polystyrene derivative to produce **21–23** [55].

These polymer metallocatalysts effectively cleaved peptide bonds of myoglobin by hydrolysis. Proteolytic activity increased considerably as the catalytic group density was raised: the ratio of k_{cat}/K_m was 1:13:100 for the mono-, di-, and tetranuclear catalysts, respectively. In the degradation of myoglobin by the dinuclear catalyst (**22**), two pairs of intermediate proteins accumulated, as revealed by MALDI-TOF MS of the reaction mixture (Figure 3.11). To identify the cleavage sites leading to the formation of the protein fragments, C-terminal sequencing was carried out by incubating the intermediate proteins with carboxypeptidase A. Analysis based on MALDI-TOF MS of the products obtained by treatment with carboxypeptidase A identified the peptide linkages of Gln(91)-Ser(92) and Ala(94)-Thr(95) as the initial cleavage sites leading to the formation of the protein fragments.

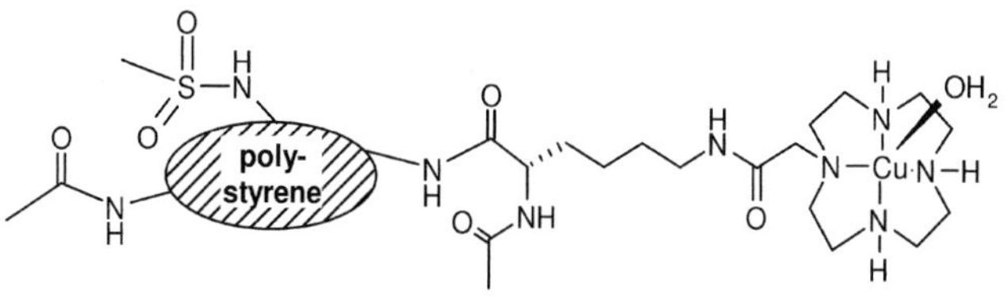

21

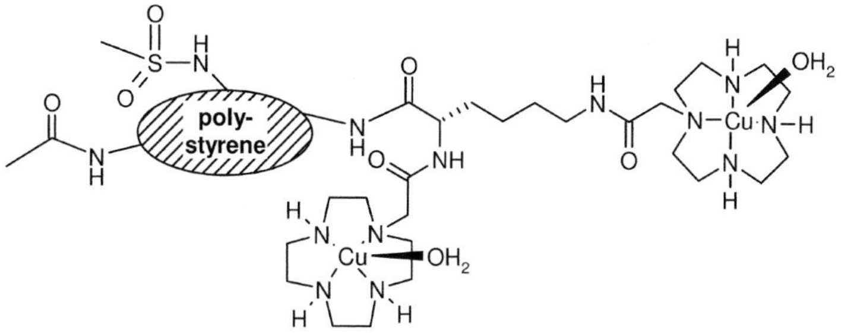

22

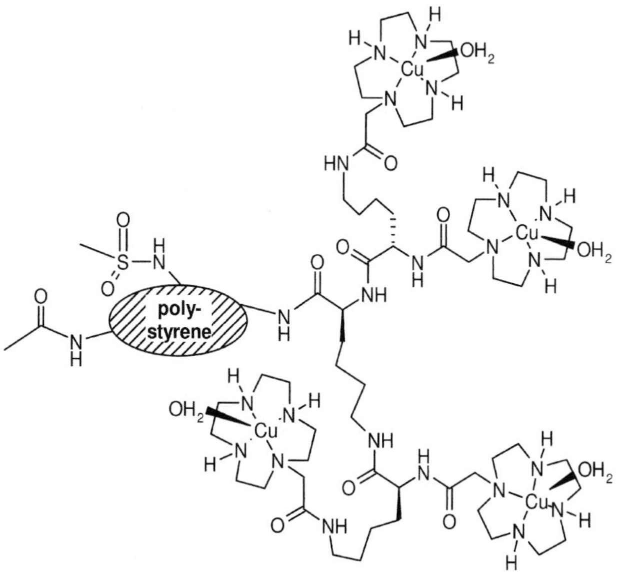

23

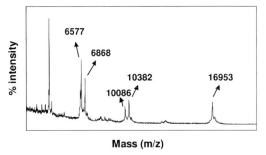

Figure 3.11 MALDI-TOF MS spectrum of degradation products obtained by incubation of myoglobin (1.2×10^{-5} M) with **22** ($C_o = 2.3 \times 10^{-3}$ M) at pH 9.0 and 50 °C for 4 h. Peaks with m/z of 6868 and 10 086 originate from cleavage of myoglobin (m/z 16 953; the heme group dissociates during measurement of MALDI-TOF MS) at Gln(91)-Ser(92), and m/z 6577 and 10 382 arose from cleavage of myoglobin at Ala(94)-Thr(95).

A molecular modeling study, based on the X-ray crystallographic structure of myoglobin, suggested that the site-selectivity was due to anchorage of one Cu(II)Cyc unit of the catalytic module to a heme carboxylate of myoglobin (Figure 3.12). The myoglobin

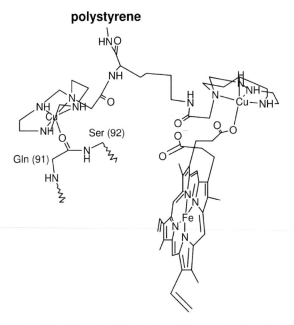

Figure 3.12 Schematic of the structure of the complex formed between **22** and myoglobin leading to cleavage of myoglobin at Gln(91)-Ser(92). The picture was obtained through optimization of the structure of the active site of **22** by molecular modeling programs with the structure of myoglobin frozen as that in the crystal.

molecule might have some structural features that guide foreign molecules to the heme pocket. The heme carboxylate, located at the gate of the entrance into the heme pocket, may serve as the initial contact point for catalysis. With a Cu(II)Cyc moiety of Cu(II)B-PS$^{MS/Ac}$ anchored to the heme carboxylate, the other Cu(II)Cyc can interact with the peptide group of Gln(91)-Ser(92) or Ala(94)-Thr(95) without significant steric strain, as examined with molecular mechanics computer programs.

3.3.4
Adducts giving Nuclease Activity to Polymers

Catalytic hydrolysis of DNA is much more challenging than that of RNA or peptides in view of the ca. 10^8-fold higher stability of DNA [56]. Most known synthetic catalysts for DNA hydrolysis are metal complexes [57–59]. Although several metal complexes promote cleavage of supercoiled and open circular DNA and single- or double-stranded oligodeoxyribonucleotides, hydrolytic cleavage of linear double-stranded polydeoxyribonucleotides by metal complexes has seldom been observed. One successful report is that for the dicerium complex **24**, which hydrolyzes a long linear DNA duplex [60].

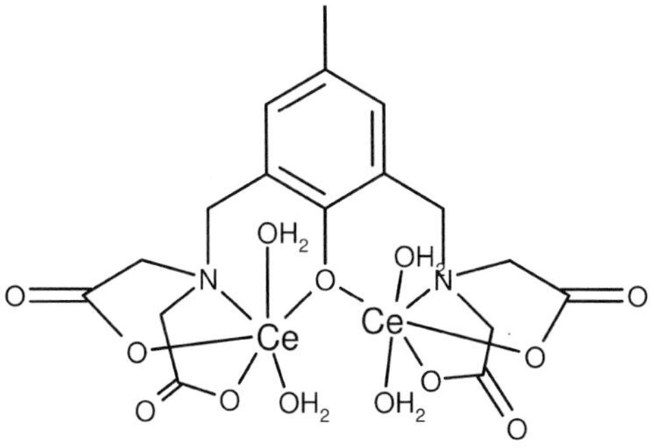

24

The Co(III)-complex of cyclen, Co(III)Cyc, is one of the most effective synthetic catalysts discovered so far for the hydrolysis of supercoiled DNAs [59]. The hydrolytic nature of DNA cleavage by the Co(III) complexes of polyamines including cyclen has been well documented [57, 58]. The mechanism illustrated in **25** has been proposed [57] for the catalytic action of the Co(III) complexes. Given the remarkable enhancement of proteolytic activity of Cu(II)Cyc upon attachment to PCD [49], we tested the activity of Co(III)Cyc in phosphodiester hydrolysis to see if it is also enhanced greatly upon attachment to PCD derivatives [61, 62].

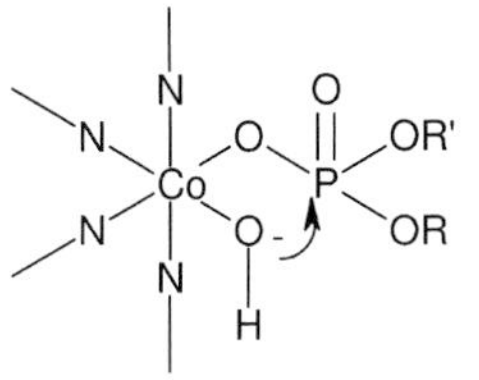

25

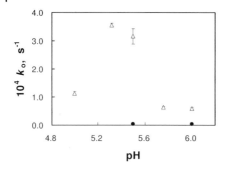

Figure 3.13 Dependence on pH of k_o for cleavage of the linearized pUC18 DNA by **26** (Δ; $C_o = 0.48$ mM) or by CoCyc (●) ($C_o = 1.0$ mM) at 25 °C.

Plasmid pUC18 DNA (2686 base pairs) was used as the supercoiled DNA substrate, whereas the linearized form of plasmid pUC18 DNA prepared by EcoR I digest was used as the linear DNA substrate. The disappearance of the supercoiled and the linear DNA substrates during incubation with several forms of PCD-supported Co(III)Cyc, such as **26**, was monitored by agarose gel electrophoresis. Kinetic data were collected in the presence of excess catalyst at 4 °C for the supercoiled DNA and at 25 °C for the linear DNA. Supercoiled DNA exhibited faster rates than linear DNA for both Co(III)-Cyc and PCD-supported Co(III)Cyc. Half-lives as short as 40 min at 4 °C and 30 min at 25 °C were observed for hydrolysis of the supercoiled and linear DNA, respectively, when catalyzed by **26** (Figure 3.13). Comparison of the rate constants measured for Co(III)Cyc and for the PCD-based Co(III)Cyc at the same catalyst concentration revealed that the Co(III)Cyc reactivity is enhanced upon attachment to PCD by more than 200 times towards the supercoiled DNA [61] and by at least 150 times towards the linear DNA [62]. Considering that only the Co(III)Cyc moieties exposed on the surface of PCD can attack the DNA, the normalized degree of activation should be much greater than 150–200-fold. As indicated by **27**, even among those exposed on the resin surface only a minor fraction of the Co(III)Cyc moieties can act as the catalytic groups. Then, the normalized degree of activation should be considerably greater than that estimated above.

Dicerium complex **24** took 24 h at 37 °C or 5 h at 55 °C to yield hydrolysis products of a 192-base pair DNA [60]. Conversely, PCD-based Co(III)Cyc derivatives degraded the linear DNA duplex into small fragments in a few hours at 25 °C. The facile DNA hydrolysis by the PCD-based Co(III)Cyc was due to the activation of Co(III)Cyc upon attachment to PCD rather than to cooperation among two or more catalytic centers or facilitation of complex formation between DNA and the PCD-supported catalyst.

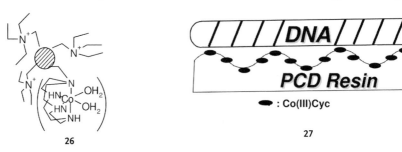

3.4
Prospectives

The last few decades has seen substantial progress in the fabrication of synthetic polymers with biocatalytic properties. A range of polymers has been examined as structural frameworks for the attachment of catalytic groups. For homogeneous catalysts, highly branched polyethylenimines have proved particularly versatile. Modified polystyrenes have served well as foundations for heterogeneous catalysts.

Increasingly sophisticated catalytic domains have been synthesized and used as adducts to the framework polymers. These synthetic macromolecules show substantially enhanced catalytic effects on hydrolytic reactions, decarboxylation, Schiff base hydrolysis, aromatic nucleophilic substitution, and oxidation [63–69]. Several of these synthetic polymers are effective peptidases and nucleases.

The local environments of catalytic domains need further modification to achieve steric and enantio-selectivity in substrates. One might also alter the polymer framework, which exerts a global influence on the local domains of active sites. Enzymes have catalytic sites attached to polypeptide framework, constituted of successive peptide linkages. Within a protein the concentration of amide groups is 12–15 M [70]. An amide group has a dipole moment [71] of 3.8 Debye units, which is larger than that of a water molecule (1.8 Debye units). Thus the protein framework provides a pervasive highly polar environment. A similar environment could be created in polyethylenimines, for example, by attaching polypeptide chains to some of the primary and secondary nitrogen loci of the polymer. These should exert a global polar influence on the catalytic domains.

Over a period of about 10^9 years (one gigaennium) natural selection has led to enzymes with remarkable catalytic properties. Perhaps, in the early decades of the 3rd millennium, directed selection will produce equally effective and versatile polymer biocatalysts.

References

1 A. J. Brown, *J. Chem. Soc.* **1902**, *81*, 373.
2 V. Henry **1903**, *Lois Générales de l' action des Diastases*, Hermann, France.
3 L. Michaelis, M. L. Menten, *Biochem. Z.* **1913**, *49*, 333.
4 I. M. Klotz **1997**, *Ligand–Receptor Energetics: A Guide for the Perplexed*, John Wiley and Sons, USA.
5 H. Bennhold, R. Z. Schubert. *Ges. Exp. Med.* **1943**, *113*, 722.
6 C. Wunderly, *Arzneim.-Forsch.* **1950**, *4*, 29.
7 U. P. Strauss, E. G. Jackson, *J. Polym. Sci.* **1951**, *6*, 649.
8 W. Scholtan, *Makromol. Chem.* **1953**, *11*, 131.
9 S. Saito, *Kolloid Z.* **1957**, *154*, 19.
10 I. M. Klotz, V. H. Stryker, *J. Am. Chem. Soc.* **1960**, *82*, 5169.
11 P. Molyneux, H. P. Frank, *J. Am. Chem. Soc.* **1961**, *83*, 3169.
12 I. M. Klotz, K. Shikama, *Arch. Biochem. Biophys.* **1968**, *123*, 551.
13 L. E. Davis **1968**, in *Water Soluble Resins*, ed. R. L. Davidson, M. Sittig, Reinhold, USA, 216.
14 I. M. Klotz, G. P. Royer, A. R. Sloniewsky, *Biochemistry*, **1969**, *8*, 4752.
15 G. P. Royer, I. M. Klotz, *J. Am. Chem. Soc.* **1969**, *91*, 5885.
16 I. M. Klotz, G. P. Royer, I. S. Scarpa, *Proc. Nat. Acad. Sci. U.S.A.* **1971**, *68*, 263.
17 T. C. Bruice, G. L. Schmir, *J. Am. Chem. Soc.* **1957**, *79*, 1663.
18 E. Katchalski, G. D. Fasman, E. Simons, E. R. Blout, F. R. N. Gurd, W. L. Koltun, *Arch. Biochem. Biophys.* **1960**, *88*, 361.

19 T. Kunitake, S. Shinkai, *J. Am. Chem. Soc.* 1971, 93, 4247.

20 T. Kunitake, Y. Okahata, *Macromolecules* 1976, 9, 15.

21 T. Kunitake, Y. Okahata, *J. Am. Chem. Soc.* 1976, 98, 7793.

22 T. Kunitake, Y. Okahata, *Adv. Polym. Sci.* 1976, 20, 217.

23 I. S. Scarpa, H. C. Kiefer, I. M. Klotz, *Intra-Sci. Chem. Rep.* 1974, 8, 45.

24 Y. Birk, I. M. Klotz, *Bioorg. Chem.* 1971, 1, 275.

25 T. Murachi, K. Okamura, *J. Polym. Sci. Polym. Lett. Ed.* 1976, 14, 361.

26 I. M. Klotz, E. N. Drake, M. Sisido, *Bioorg. Chem.* 1981, 10, 63.

27 M. Nango, E. P. Gamson, I. M. Klotz, *J. Polym. Sci. Polym. Chem. Ed.* 1979, 17, 1557.

28 M. A. Hierl, E. P. Gamson, I. M. Klotz, *J. Am. Chem. Soc.* 1979, 101, 6020.

29 S. Shinkai, H. Tsuji, Y. Harta, O. Manabe, *Bull. Chem. Soc. Jpn.* 1981, 54, 631.

30 E. J. Delaney, L. E. Wood, I. M. Klotz, *J. Am. Chem. Soc.* 1982, 104, 799.

31 W. Steglich, G. Höfle, *Angew. Chem. Int. Ed. Engl.* 1969, 8, 981.

32 G. Höfle, W. Steglich, H. Vorbrüggen, *Angew. Chem. Int. Ed. Engl.* 1978, 17, 569.

33 A. Radzicka, R. Wolfenden, *J. Am. Chem. Soc.* 1996, 118, 6105.

34 R. S. Smith, D. E. Hansen, *J. Am. Chem. Soc.* 1998, 120, 8910.

35 J. Suh, *Acc. Chem. Res.* 2003, 36, 562.

36 J. Suh, S. S. Hah, *J. Am. Chem. Soc.* 1998, 120, 10 088.

37 J. Suh, S. H. Lee, *J. Org. Chem.* 1998, 63, 1519.

38 A. Kirby, *Adv. Phys. Org. Chem.* 1980, 17, 183.

39 J. Suh, *Acc. Chem. Res.* 1992, 25, 273.

40 N. H. Williams, J. Chin, *Chem. Commun.* 1996, 131.

41 S. Liu, A. D. Hamilton, *Bioorg. Med. Chem. Lett.* 1997, 7, 1779.

42 P. Molenveld, J. F. J. Engbersen, H. Kooijman, A. L. Speck, D. N. Reinhoudt, *J. Am. Chem. Soc.* 1998, 120, 6726.

43 J. Suh, S. H. Hong, *J. Am. Chem. Soc.* 1998, 120, 12 545.

44 J. Suh, S. Oh, *J. Org. Chem.* 2000, 65, 7534.

45 H. Kim, H. Paik, M.-s. Kim, Y.-S. Chung, J. Suh, *Bioorg. Med. Chem. Lett.* 2002, 12, 2557.

46 M. L. Bender, J. A. Reinstein, M. S. Silver, R. Mikulak, *J. Am. Chem. Soc.* 1965, 87, 4545.

47 H. Kim, M.-s. Kim, H. Paik, Y.-S. Chung, I. S. Hong, J. Suh, *Bioorg. Med. Chem. Lett.* 2002, 12, 3247.

48 S. Oh, W. Chang, J. Suh, *Bioorg. Med. Chem. Lett.* 2001, 11, 1469.

49 B.-B. Jang, K. P. Lee, D. H. Min, J. Suh, *J. Am. Chem. Soc.* 1998, 120, 12 008.

50 J. Suh, S.-J. Moon, *Inorg. Chem.* 2001, 40, 4890.

51 S.-J. Moon, J. W. Jeon, H. Kim, M. P. Suh, J. Suh, *J. Am. Chem. Soc.* 2000, 122, 7742.

52 J. Suh, W. Cho, S. Chung, *J. Am. Chem. Soc.* 1985, 107, 4530.

53 J. Suh, S. B. Hong, S. Chung, *J. Biol. Chem.* 1986, 108, 7112.

54 C. Gao, B. J. Lavey, C.-H. L. Lo, A. Datta, P. Wentworth, Jr., K. D. Janda, *J. Am. Chem. Soc.* 1998, 120, 2211.

55 C. E. Yoo, P. S. Chae, J. E. Kim, E. J. Jeong, J. Suh, *J. Am. Chem. Soc.* 2003, 125, 14580.

56 N. H. Williams, B. Takasaki, M. Wall, J. Chin, *Acc. Chem. Res.* 1999, 32, 485.

57 N. E. Dixon, R. J. Geue, J. N. Lambert, S. Moghaddas, D. A. Pearce, A. M. Sargeson, *Chem. Commun.* 1996, 1287.

58 R. Hettich, J.-J. Schneider, *J. Chem. Soc., Perkin Trans.* 1997, 2, 2069.

59 R. Hettich, J.-J. Schneider, *J. Am. Chem. Soc.* 1997, 119, 5638.

60 M. E. Branum, A. K. Tipton, S. Zhu, L. Que, Jr. *J. Am. Chem. Soc.* 2001, 123, 1898.

61 C. S. Jeung, C. H. Kim, K. Min, S. W. Suh, J. Suh, *Bioorg. Med. Chem. Lett.* 2001, 11, 2401.

62 C. S. Jeung, J. B. Song, Y. H. Kim, J. Suh, *Bioorg. Med. Chem. Lett.* 2001, 11, 3061.

63 J. Suh, I. S. Scarpa, I. M. Klotz, *J. Am. Chem. Soc.* 1976, 98, 7060.

64 F. Hollfelder, A. J. Kirby, D. S. Tawfik, *J. Am. Chem. Soc.* 1997, 119, 9578.

65 W. J. Spetnagel, I. M. Klotz, *J. Am. Chem. Soc.* 1976, 98, 8199.

66 J. Suh, I. M. Klotz, *J. Polym. Sci. Pol. Chem. Ed.* 1978, 16, 1943.

67 J. Suh, I. M. Klotz, *Bioorg. Chem.* 1977, 6, 165.

68 W. J. Spetnagel, I. M. Klotz, *Biopolymers*, 1978, 17, 1657.

69 J. Suh, I. M. Klotz, *J. Am. Chem. Soc.* 1984, 106, 2373.

70 I. M. Klotz, *Brookhaven Symp. Biol.*, 1960, 13, 25.

71 E. Cohn, J. T. Edsall 1943, *Proteins, Amino Acids and Peptides*, Reinhold Publishing Co., USA, 144.

4

Mimicking Enzymes with Antibodies

Donald Hilvert

4.1
Introduction

Enzymes have fascinated chemists and biologists for over a century, as few man-made agents match the speed or selectivity of these natural catalysts. Although chemical synthesis and recombinant technologies have made many enzymes available for medical and biotechnological applications, the design of artificial protein catalysts for tasks unimagined in biology remains a challenging undertaking.

Diverse strategies for creating new enzymes have been explored. One of the most versatile exploits the mammalian immune system's ability to produce high-affinity receptor molecules, called antibodies, for a nearly limitless range of ligands. Immune recognition is based on shape and chemical complementarity between antibodies and their antigens [1]. As a consequence, immunization with appropriately designed template molecules can afford tailored antibody binding sites that exhibit many of the properties of authentic enzymes, including rate accelerations, substrate specificity, and regio- and stereoselectivity.

Antibody catalysts have been created for many classes of reaction [2, 3]. In addition to simple model reactions, transformations for which natural enzymes are unavailable have been successfully promoted. From a practical standpoint, the exacting control of reaction pathway and absolute stereochemistry that can be achieved with these agents is particularly notable. Because genetic and structural information is generally readily available, these catalysts are also valuable tools for studying how natural enzymes work and evolve.

This chapter surveys the significant progress made in engineering catalytic antibodies over the last two decades, using a few representative examples to illustrate key lessons and to highlight some of the challenges still facing this technology. Where possible, evolutionary and structural aspects are emphasized.

Artificial Enzymes. Edited by R. Breslow
Copyright © 2005 WILEY-VCH Verlag GmbH & Co. KGaA, Weinheim
ISBN: 3-527-31165-3

4.2
Basic Strategy

Antibodies are modular proteins belonging to the immunoglobulin superfamily. They contain two heavy and two light chain polypeptides, which are assembled at the genetic level by combining different V (= variable), D (= diversity), J (= joining), and C (= constant) gene segments. The resulting germline repertoire encompasses approximately 10^8 different sequences [4]. Upon immunization, individual germline antibodies are selected on the basis of their affinity for the antigen, and then subsequently refined by somatic mutation and further rounds of affinity selection. In this way, tight binding receptors for virtually any molecule can be evolved within a few weeks or months.

At the protein level, antigen recognition is mediated by two identical binding pockets formed by six peptide loops extending from the core β-barrel structure of the N-terminal antibody domains [1]. These loops, also called complementarity determining regions (CDRs), are highly variable with respect to length and amino acid sequence. Combinatorial association of these hypervariable elements makes recognition of the universe of antigens possible.

Structural studies show that antibody binding sites have the dimensions of a typical enzyme active site [1]. Moreover, they tend to be rich in tryptophans and tyrosines; asparagines and histidines are also frequently found [5]. These residues are useful for constructing generic binding sites that recognize chemically diverse ligands through hydrophobic and hydrogen-bonding interactions. Electrostatic complementarity between protein and ligand, achieved through judicious placement of charged residues within the binding pocket, is another common feature.

Given the presence of potentially reactive amino acid side chains in the confined space of the combining site, it is not surprising in retrospect that antibodies, like serum albumins [6], might chemically modify the ligands they bind. In fact, stoichiometric reactions between antibodies and labile esters have been well known for some time [7], and germline antibodies have been found to catalyze various other transformations [8]. Moreover, recent work has shown that virtually any antibody has the potential to generate toxic oxidants capable of destroying antigens [9]. In general, however, the likelihood of finding catalysts for specific reactions of interest in the primary immunological repertoire is low, and the activities observed tend to be modest.

The probability of identifying antibodies with catalytic properties can be increased if a molecule carrying chemical information about a particular reaction mechanism is used to induce the immune response. As originally proposed by Jencks [10], a stable analog of the transition state of the target reaction would be an ideal immunogen. Antibodies that evolve to bind such a compound tightly would also be expected to stabilize the true transition state and thus speed the conversion of substrate into product. The feasibility of this approach is now well established: antibody catalysts for diverse transformations, extending from hydrolytic reactions to pericyclic processes, have been elicited in response to transition state analogs [2, 3]. These catalysts exhibit Michaelis–Menten kinetics and achieve significant rate accelerations, usually between

10^3- and 10^5-fold over background. Although their activity is typically many orders of magnitude below that of comparable enzymes, they generally exhibit programmable and enzyme-like regio- and stereoselectivity.

4.3
Evolution of Binding Affinity and Catalytic Efficiency

Acyl transfer has been the most thoroughly investigated process with respect to antibody catalysis. Attention paid to such systems reflects the broad utility of hydrolytic enzymes, on the one hand, and the availability of excellent transition state analogs, on the other.

Phosphonates and phosphonamidates, which mimic the tetrahedral geometry and anionic character of the transition state for hydrolysis, have proven especially reliable as haptens. Antibodies generated against such compounds readily promote the cleavage of esters and, in a few cases, amides. High levels of stereospecificity are attainable even at chiral centers remote from the reaction site [11].

Antibody 48G7, elicited with *p*-nitrophenyl phosphonate **1**, is a representative esterase [12]. It accelerates the hydrolysis of activated ester **2a** and carbonate **2b** by factors of >10^4 (Scheme 4.1). Detailed study of its structure and immunological origins has provided valuable insight into the mechanism and evolution of catalysis.

As programmed by the phosphonate hapten, 48G7 is a relatively simple catalyst that facilitates direct hydroxide attack on the scissile carbonyl of its ester substrate. The antibody–hapten complex shows **1** bound in an extended conformation with the aryl leaving group buried deeply in a hydrophobic cleft [13]. The negatively charged phosphonate sits near the entrance of the pocket, where it forms hydrogen bonds with the side chains of Tyr^{H33}, His^{H35}, Arg^{L96}, and the backbone amide of Tyr^{H96}. This collection of polar residues is reminiscent of the oxyanion hole in serine proteases, and presumably stabilizes the oxyanionic transition states of the hydrolysis reaction through analogous hydrogen-bonding and ionic interactions.

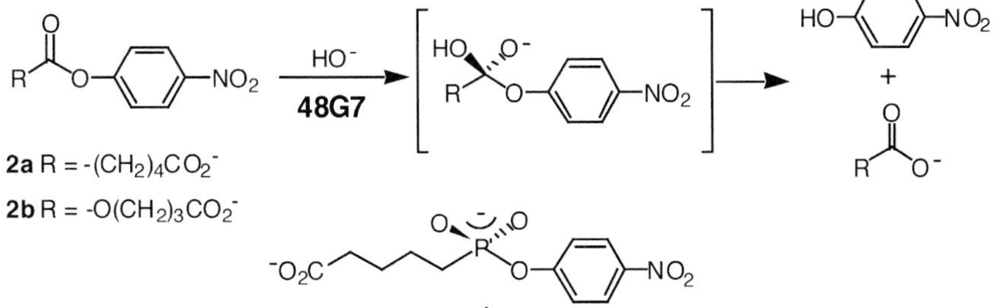

2a R = -$(CH_2)_4CO_2^-$
2b R = -$O(CH_2)_3CO_2^-$

1

Scheme 4.1 Hydrolysis of activated aryl esters **2a** and carbonates **2b** proceeds via an anionic, tetrahedral intermediate (in square brackets). Hydrolytic antibody 48G7 was elicited with an aryl phosphonate derivative **1** that mimics this high energy species and its flanking transition states.

The germline precursor of 48G7 was isolated from the primary immunological repertoire [14]. It also accelerates ester hydrolysis, yet is 20-fold less efficient than the mature antibody and binds the hapten 30 000-fold less tightly. During affinity maturation, nine somatic mutations were introduced into the antibody scaffold. Although none of the altered residues directly contacts bound **1**, the substitutions appear to organize the pocket for binding and catalysis by limiting side-chain and backbone flexibility inherent in the germline protein [15]. Thus, ligand binding to germline 48G7 induces significant structural reorganization, whereas the free and ligand-bound forms of the mature antibody are very similar. The germline and mature antibodies also display some differences in the way they recognize the hapten and, by extension, the hydrolytic transition state. For example, because of an altered conformation for the CDR H1 loop, the aryl leaving group adopts different orientations at the two active sites. More importantly, due to conformational constraints, Tyr^{H33} cannot provide a hydrogen bond to the anionic transition state in the germline complex, resulting in reduced catalytic efficiency.

Parallel evolution of hapten binding and catalytic activity has been observed in several catalytic antibodies in addition to 43G7 [8, 16], illustrating the value of immunization as a means of optimizing the catalytic potential of the antibody combining site. Nevertheless, the correlation between binding and catalysis is necessarily approximate because stable analogs can never perfectly mimic the stereoelectronic features of a high energy transition state. Sometimes affinity maturation even causes a loss of activity, as seen in antibody AZ-28 [17], which catalyzes an oxy-Cope rearrangement. Mature AZ-28 binds the transition state analog 40-times more tightly than its germline precursor, but it is 30-fold less efficient as a catalyst. This inverse relationship has been attributed to structural differences between the transition state analog and the transition state, and differences in the flexibility of the mature and germline antibodies [18].

Because only a fraction of the antibodies induced in response to any given transition state analog typically exhibits catalytic activity, screening is an essential step in any catalytic antibody experiment. Consequently, much effort has been invested in the development of sensitive high-throughput assays [19]. For the same reason, access to large and diverse populations of hapten-binders is desirable. With respect to the latter point, autoimmune mice yield unusually large numbers of esterolytic antibodies [20], and may usefully expand the repertoire of catalytic clones elicited by other transition state analogues.

4.4
Importance of a Good Fit

The ability of enzymes to preorganize reactants for reaction contributes significantly to their efficacy [21]. The excellent shape complementarity observed in antibody–antigen complexes suggests that antibody binding energy could be exploited for this purpose. Antibody catalysts have been developed for bimolecular Diels–Alder reactions to test this supposition.

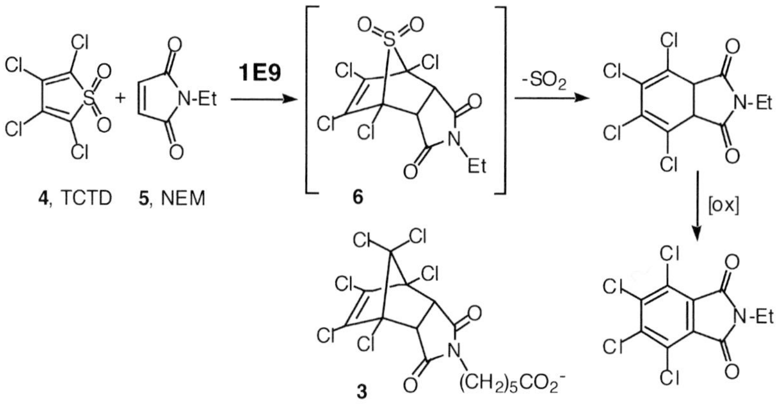

Scheme 4.2 Diels–Alder cycloaddition of tetrachlorothiophene dioxide (**4**) and N-ethyl maleimide (**5**) yields a high energy intermediate (**6**)
that decomposes spontaneously by elimination of sulfur dioxide. The
initially formed product is subsequently oxidized in situ. Transition
states for cycloaddition and chelotropic SO_2 elimination closely resemble the hexachloronorbornene derivative **3** used as a hapten to elicit
antibody 1E9.

Antibodies raised against neutral bicyclic compounds that mimic the boat-like geometry of the Diels–Alder transition state have achieved significant catalytic effects, including
large rate accelerations, multiple turnovers, and control over both reaction pathway and
absolute stereochemistry [2]. For example, the hexachloronorbornene derivative **3** is an
excellent mimic of the transition state for the [4+2] cycloaddition between tetrachlorothiophene dioxide (**4**) and N-ethylmaleimide (**5**) (Scheme 4.2) [22]. Antibodies that bind this
compound catalyze the Diels–Alder reaction efficiently and experience minimal product
inhibition because large structural changes ensue when the initially formed product (**6**)
eliminates sulfur dioxide and becomes oxidized [23]. One of these antibodies, 1E9, attains
an effective molarity (EM) in excess of 1000 M [22]. The EM represents the diene (or
dienophile) concentration needed to convert the uncatalyzed reaction into a pseudo-un-
imolecular process as fast as the reaction at the antibody active site. As such, it is a measure of the advantage the antibody derives from preorganizing the substrates for reaction.
By comparison, EMs seldom exceed 20 M for other antibodies [24–26], ribozymes [27],
and even natural enzymes [28] that promote Diels–Alder reactions.

Structural studies indicate that 1E9's efficiency results from an unusually snug fit
between the protein and the cycloaddition transition state [22]. Approximately 86% of
the hapten is buried upon binding, with no interfacial cavities discernable between
protein and ligand. This tight packing is achieved through extensive van der Waals
contacts and π-stacking interactions. In addition, the side chain of Asn[H35] at the
base of the pocket provides a hydrogen bond to the buried succinimide carbonyl group.
Such interactions are ideally suited to preorganize the diene and dienophile for pro-
ductive reaction, and to stabilize the transition state enthalpically [29].

The immunological origins of 1E9 have been traced to a family of highly restricted
germline antibodies that bind diverse hydrophobic ligands [30]. A precise fit to the

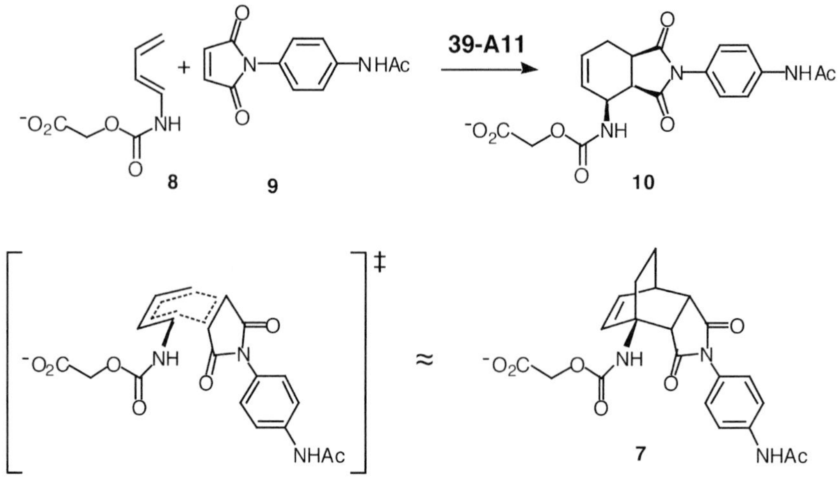

Scheme 4.3 Antibody 39-A11 catalyzes a Diels–Alder reaction between an electron-rich acyclic diene (**8**) and an N-aryl maleimide (**9**). It was elicited with the bicyclo[2.2.2]octene hapten **7**. The ethano bridge locks the cyclohexene ring into the requisite boat conformation but has no counterpart in the substrates or transition state.

hexachloronorbornene hapten was achieved by introducing two rare somatic mutations into the germline scaffold, one at position L89 in the light chain (Ser^{L89}Phe) and another at H47 in the heavy chain (Trp^{H47}Leu) [22]. These seemingly subtle substitutions, coupled with minor adjustments in CDR H3 residues, were sufficient to convert a relatively non-specific, catalytically inactive germline antibody into a potent catalyst.

This same family of germline antibodies has also given rise to a second Diels–Alderase, antibody 39-A11. In this case, the substituted bicyclo[2.2.2]octene derivative **7** served as hapten [24]. This antibody accelerates the reaction between an electron-rich acyclic diene (**8**) and an N-aryl maleimide (**9**) to give a cyclohexene derivative (**10**) (Scheme 4.3), but catalysis is relatively inefficient as judged by an EM of 0.35 M. Examination of the 39-A11 structure points to an explanation for the low efficiency: the large bicyclooctene unit used to mimic the reacting [4+2] system is relatively poorly packed by the protein [31]. Most binding interactions in the complex are directed instead to the aryl group shared by the substrate and transition state. The binding pocket in which the cycloaddition takes place is simply too large to restrict all the degrees of freedom available to the diene substrate, so low rates result. By improving packing interactions with the cycloaddition transition state through introduction of large aromatic groups at positions L91 and L96, it was possible to increase k_{cat} by 5- to 10-fold [32].

These examples illustrate the important interplay between binding energy and catalysis, underscoring the utility of complementary packing interactions at the transition state. For 1E9, the mechanistic information contained in the transition state analog is effectively mirrored in the induced binding pocket, allowing efficient catalysis. Even so, 1E9 is not an evolutionary optimum with respect to activity. Despite the seemingly perfect shape complementarity between protein and transition state analog, replacement of Met^{H100b} at the floor of the pocket by phenylalanine increased the affinity for **3** by a factor of two and augmented k_{cat} 7-fold [33]. Even higher activity is likely to be attainable if the affinity of the antibody for the transition state (analog) can be further increased.

Although nanomolar affinities are generally sufficient for the purposes of the immune system, they are inadequate for high catalytic activity. Enzymes are believed to bind transition states many orders of magnitude more tightly than 1E9 binds the hexachloronorbornene **3** (K_d = 0.1 nM [34]). In effect, with respect to catalysis, affinity maturation stops too soon. In principle, it should be possible to continue the evolution of these first generation catalysts in vitro by subjecting them to random mutagenesis, and screening the resulting libraries for tighter ligand binding or directly for higher activity. Nanomolar binders have been converted into receptors with picomolar to femtomolar affinities in this way [35]. If similar effects can be achieved with 1E9, a significant fraction of the binding energy gained may be manifest as enhanced catalytic efficiency.

4.5
General Acid–Base Catalysis

Highly evolved enzymes often possess sophisticated arrays of functional groups that they use to promote otherwise difficult reactions. Although catalytically useful acids, bases, or nucleophiles can arise by chance in the immunoglobulin pocket during evolution of the immune response to transition state analogs, other measures are generally needed to elicit reactive residues reliably and systematically. Charge complementarity between antibody and antigen has been extensively exploited for this purpose. For example, positively charged ammonium salts have yielded various antibodies that utilize carboxyl groups to promote eliminations, enolizations, hydrolytic reactions and isomerizations [2].

Antibody catalysts for converting benzisoxazoles (**11**) into salicylonitriles (**12**) (Scheme 4.4), a well-characterized E2 elimination that is sensitive to base strength and solvent microenvironment [36], illustrate the strategic use of haptenic charge. The cationic 2-aminobenzimidazolium derivative **13**, which mimics the overall geometry of the elimination transition state and bears little resemblance to the reaction product, was employed as a hapten [37]. Antibody 34E4, which binds **13** tightly ($K_d \approx$ 1 nM), catalyzes the target reaction with more than 10^3 turnovers per active site and a rate acceleration of 10^6 over background. A glutamate at H50 serves as the catalytic base [38, 39]. The effective molarity for this residue is >10 000 M [37, 39], an exceptionally high value relative to EMs obtained in intramolecular model systems for general base catalysis [40].

The efficiency of 34E4 likely derives from a combination of factors [38, 41]. Medium effects are certainly relevant. The carboxylate base is activated by placement in a relatively hydrophobic environment, as evidenced by its elevated pK_a of 6 [37]. The charge delocalized transition state is probably also stabilized by dispersive interactions with

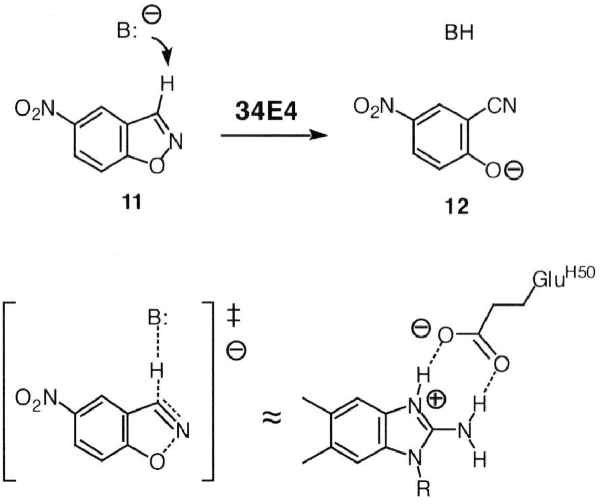

Scheme 4.4 Base-catalyzed elimination of 5-nitrobenzisoxazole (**11**) to give salicylonitrile **12** is catalyzed by antibody 34E4. Haptenic 2-aminobenzimidazolium derivative **13** induced a complementary glutamate at H50 in the binding pocket that serves as base in the catalytic reaction.

the aromatic residues that sandwich the planar benzimidazolium in the antibody–hapten structure [42]. In addition, the high degree of positional ordering imposed by the antibody active site likely contributes significantly to catalytic efficiency. Aromatic residues lining the pocket clamp the hapten in an orientation that allows formation of a bidentate hydrogen bond with the side chain of the carboxylate base, Glu^{H50} [42]. The catalytic carboxylate, itself locked in place by hydrogen bonds with the side chain of Asn^{H58} and a fixed water molecule, is optimally situated to abstract a proton from a benzisoxazole occupying the ligand binding slot.

Replacement of the catalytic glutamate by aspartate destabilizes the transition state for elimination by 2.1 kcal mol^{-1} [39]. Because the pK_a of the carboxylate in the aspartate mutant is the same as in the parent antibody, loss of activity is directly attributable to suboptimal positioning of the base relative to substrate rather than to a medium effect. This conclusion is supported by X-ray data for the mutant showing that the aspartate carboxyl group is a little farther from the ligand than the original glutamate and adopts an altered orientation that is stereoelectronically less suited for proton abstraction [42]. The advantage the antibody derives from covalently fixing its base at the active site is further underscored by findings that high formate and acetate concentrations cannot rescue the inactive variants $Glu^{H50}Ala$ and $Glu^{H50}Gly$, which lack an active site base of their own [39].

Comparison of 34E4 with a less proficient catalyst shows that merely positioning a carboxylate in a hydrophobic binding pocket does not result in efficient general base catalysis. Antibody 4B2, generated against cationic amidinium salt **14** (Scheme 4.5),

Scheme 4.5 Cationic iminium ion **14** was used to generate antibody 4B2, which has an active site carboxylate at L34 and also promotes the decomposition of benzisoxazoles shown in Scheme 4.4.

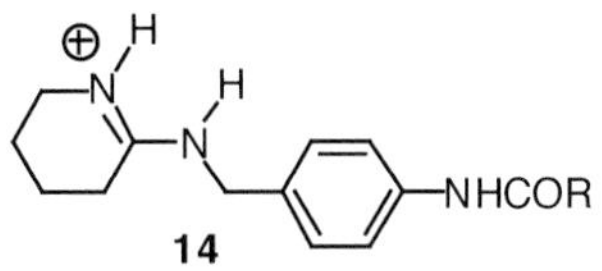

14

also accelerates the decomposition of benzisoxazoles [43], using a glutamate as the catalytic base [44]. However, despite hydrophobic surroundings and an elevated pK_a of 5.8, this residue is two to four orders of magnitude less proficient at proton abstraction than Glu^{H50} in 34E4 [39]. Bearing little resemblance to the amidinium hapten, the 5-nitrobenzisoxazole substrate presumably binds at the relatively spacious 4B2 binding pocket [44] in various orientations that are suboptimal for proton transfer.

The properties of 34E4 are impressive, demonstrating that functional groups can be exploited to excellent effect in antibody binding pockets. In fact, as judged by its EM, the catalytic carboxylate appears to be as effective at proton abstraction as analogous groups in highly evolved natural enzymes [39, 45]. Nevertheless, the overall efficiency of the antibody is still several orders of magnitude lower than enzymes such as triose phosphate isomerase, which also promote proton abstractions. In contrast to its natural counterparts, 34E4 relies on only one functional group, namely Glu^{H50}, for catalysis. Calculations suggest that provision of an additional residue to stabilize the developing negative charge at the leaving group might augment activity by several orders of magnitude [46]. Unfortunately, initial attempts to introduce such a residue by site-directed mutagenesis have been unsuccessful [39].

Construction of active sites containing several functional groups remains a significant challenge for antibody catalysis. The likelihood of eliciting multiple catalytic residues spontaneously is vanishingly small. Likewise, engineering additional groups into existing pockets, as seen with 34E4, has met with little success. No doubt this explains a relative dearth of good antibody catalysts for energetically demanding reactions, such as amide or glycoside hydrolysis. Heterologous immunization with two different but structurally related haptens, each containing a different charged moiety, was developed to address this problem [47]. Promising results with antibody esterases were reported [48], but catalysts with truly enzyme-like activities have yet to emerge.

4.6
Covalent Catalysis

Reactive immunization is a fundamentally different approach to selecting antibody pockets that contain functional groups. This method employs mechanism-based inhibitors as haptens; these molecules react covalently with appropriately functionalized antibodies, allowing direct selection of active clones from large pools of inactive variants. When a suitable substrate is used in place of the inhibitor, reactive residues in the selected antibodies can often mediate its conversion into product.

This strategy has yielded catalysts for several reactions. For example, phosphonylating agents have been used to generate antibodies that promote ester hydrolysis by a

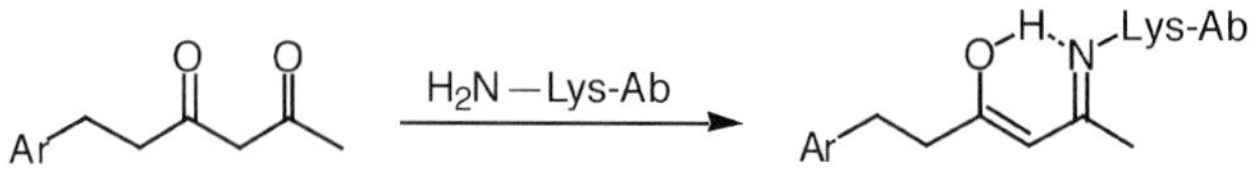

15

16, λ_{max} = 316 nm
(ε 15,000 M^{-1} cm^{-1})

Scheme 4.6 1,3-Diketone **15**, used as a hapten to raise antibody 38C2, traps the reactive, active-site amine of Lys[H93] to form a stable, chromophoric vinylogous amide.

two-step mechanism involving transient acylation of an active site nucleophile [49]. Other mechanism-based inactivators of hydrolytic enzymes have afforded antibodies with β-lactamase [50] and galactosidase activity [51]. Versatile aldolase catalysts have been created with immunogenic 1,3-diketones (**15**) [52], which trap reactive lysines via Schiff base formation, then rearrange to a more stable and readily detected vinylogous amide (**16**) (Scheme 4.6). In the presence of a β-hydroxyketone instead of the 1,3-di-ketone, analogous enamine chemistry results in a retro-aldol reaction. In the synthetic direction, a wide range of aldol condensations is catalyzed [53].

Mechanistically, the antibody aldolases resemble natural class I aldolase enzymes (Scheme 4.7) [52]. In the first step of a condensation reaction, the ε-amino group of the catalytic lysine reacts with a ketone to form a Schiff base. Deprotonation of this species yields a nucleophilic enamine, which condenses with electrophilic aldehydes in a second step to form a new carbon–carbon bond. Subsequent hydrolysis of the Schiff base releases product and regenerates the active catalyst.

Aldolase antibodies obtained by reactive immunization are notable for high activity, broad substrate specificity, and high selectivities [53]. Rate accelerations are typically in the range 10^5 to 10^7-fold over background. Although the k_{cat}/K_m values are 10^2 to 10^4 lower than those of aldolase enzymes, these are among the most efficient antibody catalysts described to date. Their efficacy is all the more notable in light of the inherently complex, multistep process they catalyze.

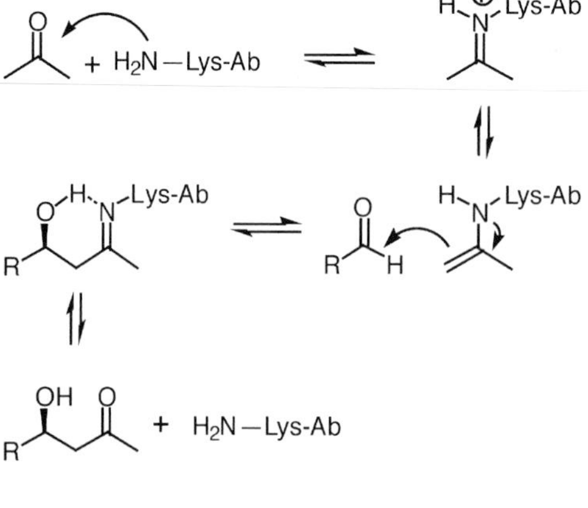

Scheme 4.7 Antibodies generated against hapten **15** promote diverse aldol condensations through enamine chemistry.

The unusually wide range of substrates accepted by these catalysts stands in contrast to the narrow substrate scope of catalysts raised against transition state analogs. This functional promiscuity has been ascribed to covalent capture of the antibody early in the process of affinity maturation [53], obviating further refinement of the binding pocket by somatic mutation. Consistent with this idea, the active site of a structurally charac-terized antibody aldolase consists of a large hydrophobic pocket, over 11 Å deep, con-taining a single lysine [53]. The hydrophobic environment around the lysine accounts for its low pK_a (between 5.5 and 6.0). Interestingly, natural aldolases adopt a completely different strategy to lower the pK_a of their catalytic lysine, exploiting electrostatic rather than hydrophobic interactions [54]. A further difference between these catalysts is the relative scarcity of other functional groups in the antibody binding pocket. For example, there is no obvious acid or base to facilitate proton transfers in the reaction. Sequestered water molecules, or possibly a tyrosine or serine residue in the pocket, may be involved, but additional structural work on antibody–ligand complexes is needed to sort this out.

Despite the absence of stereochemical information in the reactive immunogen, the aldolase antibodies promote carbon–carbon bond formation with surprisingly high selectivity. For instance, the enamine formed from acetone adds to the *si* face of var-ious aldehydes with *ee*'s in excess of 95 % [53]. In other examples, Robinson annula-tions have been carried out with high enantioselectivity [55], tertiary aldols and other compounds have been successfully resolved [56], and enantiopure intermediates have been prepared for the synthesis of various natural products [57, 58].

Reactive immunogens incorporating elements of transition state mimicry have de-livered even more efficient catalysts. Compound **17** (Scheme 4.8), for example, con-tains a tetrahedral sulfone to mimic the geometry of the acceptor site during C–C bond formation. It was used to produce antibodies that accelerate the retro-aldol reaction of **18** with a k_{cat}/K_m of 3×10^5 M^{-1} s^{-1} and a rate acceleration over background (k_{cat}/k_{uncat}) of 2×10^8 [59]. These are impressive results for a catalyst never optimized by natural selection.

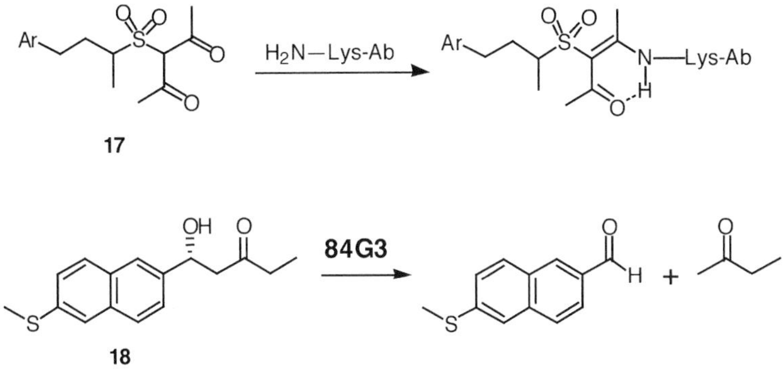

Scheme 4.8 Hapten **17**, designed to combine transition state mi-micry and reactive immunization strategies, produced an aldolase antibody (84G3) that promotes aldol reactions with typically higher rates and selectivities than antibodies raised against **15**. The retro-aldol reaction of **18** is catalyzed with notable efficiency by this antibody.

Although reactive immunization does not appear to solve the problem of creating active sites with multiple catalytic residues, this approach has the decided advantage of selecting antibodies on the basis of their ability to initiate a chemical reaction (albeit on a substrate surrogate) as opposed to tight binding to a stable hapten. The efficacy of the resulting catalysts speaks for itself.

4.7
Practical Applications

As the above examples attest, catalytic antibody technology can be a powerful and versatile approach for creating new catalysts. Not surprisingly, various practical applications have been envisaged to capitalize on this capability.

In theory, the programmable stereoselectivities of catalytic antibodies makes them well suited for asymmetric synthesis. Several such transformations have been carried out on a preparative scale. Kinetic resolution of the epothilone precursor **19** with the aldolase antibody 38C2 is instructive (Scheme 4.9) [57]. The reaction proceeds in good yield (37%) and high enantiomeric excess (90%). However, so much catalyst is needed (0.5 g of IgG antibody was used for the resolution of 0.75 g **19**) that large-scale production is likely to be impractical in many cases. As most antibody catalysts are much less efficient than the aldolases, catalyst costs will generally be appreciable.

Reactions difficult or impossible to carry out with existing methodology have been identified as another opportunity for this technology [60]. Successful antibody catalysis of normally disfavored exo rather than endo Diels–Alder cycloadditions [25], *syn* rather than *anti* eliminations [61], and 6-*endo*-tet rather than 5-*exo*-tet ring closures [62] encourage such thinking (Scheme 4.10). It is even possible to use antibody binding energy to control the energetics and interconversion of short-lived excited-state species [63]. If the low specific activities of these catalysts can be improved, many exciting applications will be realizable.

Because antibodies have long serum half-lives, they can also be used in vivo. Selective activation of prodrugs has received particular attention. For instance, inactive esters of chloramphenical [64] and 5-fluorodeoxyuridine [65] have been converted into their bioactive forms by hydrolytic antibodies, yielding sufficient antibiotic in test experiments to inhibit bacterial growth. In another example, etoposide prodrug **27** has been activated for chemotherapeutic applications by sequential aldolase-catalyzed

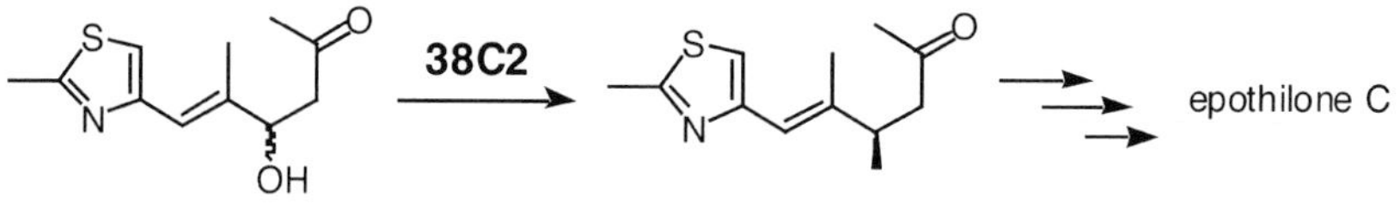

(±)-**19** 38C2 (-)-**19**, 37% yield, 90% *ee* epothilone C

Scheme 4.9 Kinetic resolution of **19**, a precursor to the natural product epothilone C, was accomplished by degrading the unwanted stereoisomer by an antibody-catalyzed retro-aldol reaction.

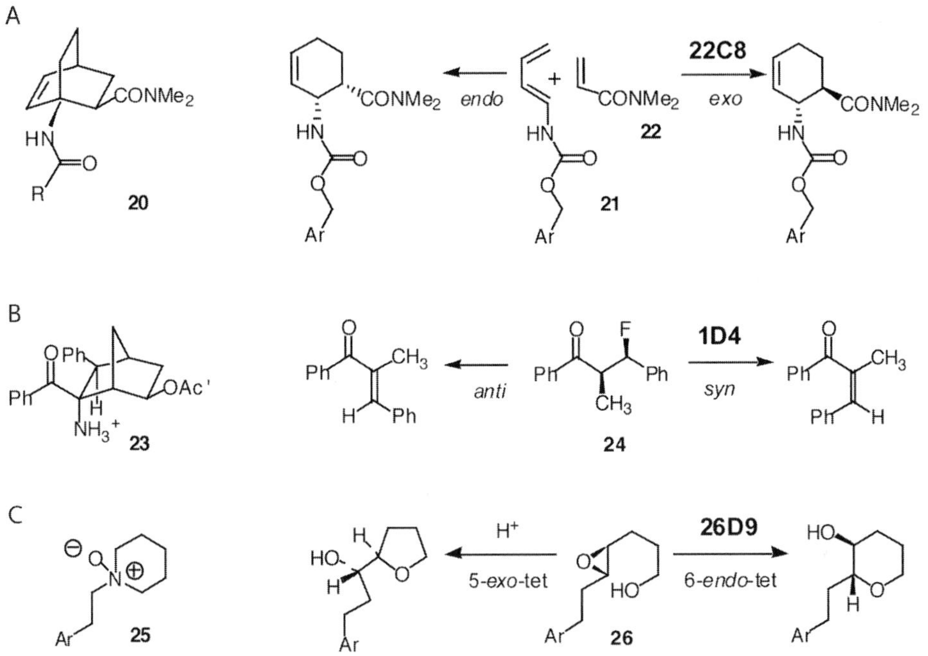

A

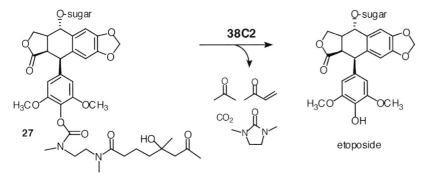

Scheme 4.10 Examples of disfavored reactions catalyzed by antibodies. (A) Antibody 22C8 was generated against the bicyclo[2.2.2]octene derivative 20 and catalyzes the exo cycloaddition of diene 21 to acrylamide 22. (B) The energetically demanding *syn* elimination of 24 is accelerated by antibody 1D4. The latter was produced by the conformationally restricted hapten 23, which contains an ammonium group to induce a catalytic base. (C) Antibody 26D9, elicited with amine oxide 25, reverses the normal preference for formation of five-membered rings in the reaction of 26 to give enantiomerically pure tetrahydropyran via the higher energy 6-endo transition state.

Scheme 4.11 Activation of a prodrug of the anticancer agent etoposide via tandem retro-aldol/retro-Michael reactions catalyzed by antibody 38C2.

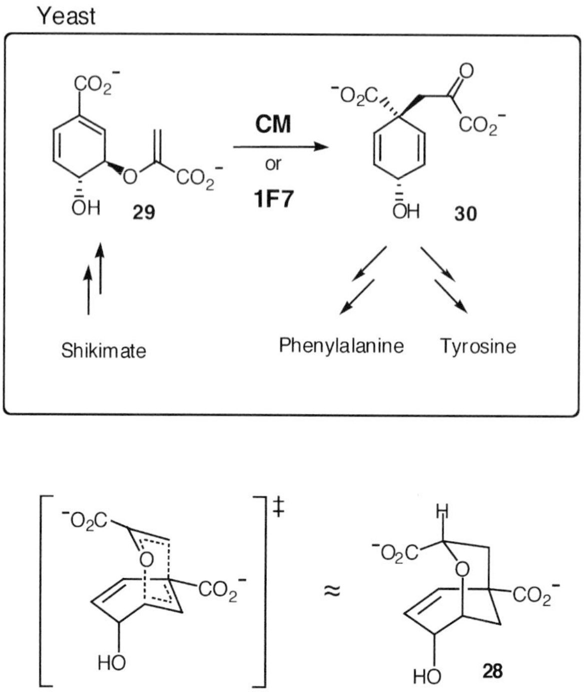

Yeast

Shikimate   Phenylalanine   Tyrosine

Scheme 4.12 Catalytic antibody 1F7 was raised against the transition state analog **28** and possesses modest chorismate mutase activity. It can complement a permissive yeast strain that is auxotrophic for phenylalanine and tyrosine by replacing the natural enzyme (CM) in the shikimate biosynthetic pathway.

retro-aldol and retro-Michael reactions (Scheme 4.11) [66]. Detoxification of otherwise toxic substances is also feasible, as shown by the protection of rats from cocaine-induced seizures and death by a cocaine-hydrolyzing antibody [67].

Successful production of immunoglobulins in microorganisms such as *Escherichia coli* and yeast opens the door to additional in vivo applications. Intracellular biosynthesis of tailored antibody catalysts could confer new properties on these organisms, such as pesticide resistance or novel biosynthetic capabilities. Experiments with an anti-**28** antibody possessing modest chorismate mutase activity [68] show that they can replace natural enzymes in cellular metabolism. When produced cytoplasmically at high concentrations in a chorismate mutase deficient yeast strain, the Fab fragment of this catalyst reconstitutes the shikimate biosynthetic pathway leading to the aromatic amino acids tyrosine and phenylalanine (Scheme 4.12), conferring a substantial growth advantage under auxotrophic conditions [69]. Although efforts to direct the evolution of this primitive antibody through cycles of mutagenesis and genetic selection have not provided substantially more active clones, they have identified the heavy chain as the minimal domain needed for function in vivo, which is consistent with the observation that this chain contributes over 90% of the contacts with the bound transition state analog [70]. If applied broadly, genetic complementation could be a powerful tool for screening the primary immunological repertoire for more active catalysts and for augmenting the activity of first-generation catalytic antibodies in directed evolution experiments.

4.8
Future Directions

By combining programmable design with the powerful selective forces of biology, catalytic antibodies merge many of the best features of synthetic and enzymatic catalysts. Nevertheless, there is a general sense that these agents have failed to fulfill their original promise. While it is relatively easy to generate moderately active catalysts for model reactions of activated substrates, production of truly efficient enzyme mimics has proven exceptionally difficult [3]. Moreover, it has not been possible to accelerate many reactions of interest to desirable rates. Hydrolysis of unactivated amides and phosphate diesters are examples of such. As a consequence, visions of site-selective proteases or restriction enzymes remain unrealized.

Reasons for these difficulties are readily enumerated [3]. Above all, the basic strategy for producing catalytic antibodies is indirect. Tight binding to an imperfect transition state analog, rather than catalysis, is the selection criterion that drives immunological evolution. Even if perfect transition state analogs were available, the nanomolar affinities routinely attainable during affinity maturation limit the extent to which a transition state can be stabilized relative to the ground state. Furthermore, the probability of identifying rare but highly active clones is low because only a tiny fraction of the immune response is experimentally accessible in a typical experiment.

The origins of inefficiency may also reside in the scaffold shared by all antibodies. Immunoglobulin structural diversity appears to be far more restricted than originally anticipated based on the vast number of germline sequences. Studies on many catalytic antibodies document the frequent recurrence of basic hapten recognition motifs. The commonalities observed in numerous anti-arylphosphonate antibodies [71], and the closely related structures of the 1E9 and 39A-11 Diels–Alderases discussed above [22], are paradigmatic. In favorable cases, such as the 1E9 antibody, the combining pocket can be molded remarkably effectively to achieve nearly perfect shape complementarity to its ligand. In general, however, these frequently selected binding pockets may be poorly suited to particular catalytic tasks or may represent local minima from which it will be difficult to evolve further.

By improving transition state analogs, refining immunization and screening protocols, and developing strategies to increase the efficiency of first-generation catalytic antibodies, it may be possible to overcome some of the hurdles in the path to enzyme-like activities. Notably, though, the immune system was originally exploited as a source of catalysts as a matter of convenience. While it is still unrivaled in biology in its ability to fashion protein receptors to virtually any natural or synthetic molecule, new methods have since emerged for creating protein libraries based on diverse scaffolds and in diverse formats. Phage display [72], cell surface display [73], ribosome display [74], and mRNA-protein fusions [75] are only some of the techniques devised. In coupling genotype and phenotype they enable in vitro selection of proteins with interesting binding or catalytic properties from vast combinatorial libraries. Genetic selection in vivo represents yet another powerful tool for enzyme (re)design that effectively harnesses the power of evolution [76].

Recent advances in computation, combined with the methods of directed evolution, should make future enzyme engineering much easier. Conversion of a ribose-binding protein into a triose phosphate isomerase mimic represents an exciting first step [77]. If the results of this study can be generalized, computation could become an effective alternative to the immune system for efficiently exploring sequence space. Such an approach would not be limited to a single scaffold, but binding pockets could be matched to the requirements of the reaction being catalyzed. It may even be possible to design new protein folds with tailored active sites de novo. Since reaction mechanism and transition state properties can be considered explicitly during the computational design process, the chances of obtaining a good experimental starting point for further development are likely to be considerably improved. As illustrated by the artificial triose phosphate isomerase example [77], random mutagenesis coupled with high-throughput screening or mutagenesis allows subsequent fine-tuning.

4.9
Outlook

Antibody catalysts are arguably the most successful enzyme mimics so far described. They can be prepared by a general strategy that takes advantage of our knowledge of reaction mechanism and physical organic chemistry. Although the rates achieved as yet are not "enzyme-like", enzymes – with millions of years of evolutionary optimization – may set an unrealistically high standard to judge synthetic catalysts that can be evolved in the immune system over a few weeks or months. This approach has already provided valuable model systems for exploring how protein binding energy can be exploited to accelerate numerous chemical transformations, including those lacking biological counterparts. Even if catalytic antibodies never achieve the broad applicability originally anticipated, the strategies employed in their productions and the lessons learned can be expected to contribute significantly to the success of future efforts to create tailored proteins with truly enzyme-like properties.

Acknowledgments

The author is indebted to the ETH Zürich, the Swiss National Science Foundation and Novartis Pharma for generous support.

References

1 D. R. Davies, E. A. Padlan, S. Sheriff, *Annu. Rev. Biochem.* **1990**, *59*, 439–473; I. A. Wilson, R. L. Stanfield, *Curr. Opin. Struct. Biol.* **1993**, *3*, 113–118.

2 P. G. Schultz, J. Yin, R. A. Lerner, *Angew. Chem. Int. Ed.* **2002**, *41*, 4427–4437; J. D. Stevenson, N. R. Thomas, *Nat. Prod. Rep.* **2000**, *17*, 535–577; D. Hilvert, *Top. Stereochem.* **1999**, *22*, 83–135; R. A. Lerner, S. J. Benkovic, P. G. Schultz, *Science* **1991**, *252*, 659–667.

3 D. Hilvert, *Annu. Rev. Biochem.* **2000**, *69*, 751–793.

4 F. W. Alt, T. K. Blackwell, G. D. Yancopoulos, *Science* **1987**, *238*, 1079–1087; K. Rajewsky, I. Förster, A. Cumano, *Science* **1987**, *238*, 1088–1094.

5 E. A. Padlan, *Proteins* **1990**, *7*, 112–124.

6 D. C. Carter, J. X. Ho, *Adv. Protein Chem.* **1994**, *45*, 153–203.

7 F. Kohen, J. B. Kim, H. R. Lindner, Z. Eshhar, B. Green, *FEBS Lett.* **1980**, *111*, 427–431; F. Kohen, J.-B. Kim, G. Barnard, H. R. Lindner, *Biochim. Biophys. Acta* **1980**, *629*, 328–337.

8 M. C. Orencia, M. A. Hanson, R. C. Stevens, *Adv. Protein Chem.* **2001**, *55*, 227–259.

9 A. D. Wentworth, L. H. Jones, P. Wentworth, K. D. Janda, R. A. Lerner, *Proc. Natl. Acad. Sci. U.S.A.* **2000**, *97*, 10930–10935; P. Wentworth, J. E. McDunn, A. D. Wentworth, C. Takeuchi, J. Nieva, T. Jones, C. Bautista, J. M. Ruedi, A. Gutierrez, K. D. Janda, B. M. Babior, A. Eschenmoser, R. A. Lerner, *Science* **2002**, *298*, 2195–2199; P. Wentworth, A. D. Wentworth, X. Y. Zhu, I. A. Wilson, K. D. Janda, A. Eschenmoser, R. A. Lerner, *Proc. Natl. Acad. Sci. U.S.A.* **2003**, *100*, 1490–1493.

10 W. P. Jencks **1969**, *Catalysis in Chemistry and Enzymology*, McGraw Hill, New York, p. 288.

11 L. J. D'Souza, B. Gigant, M. Knossow, B. S. Green, *J. Am. Chem. Soc.* **2002**, *124*, 2114–2115.

12 S. A. Lesley, P. A. Patten, P. G. Schultz, *Proc. Natl. Acad. Sci. U.S.A.* **1993**, *90*, 1160–1165.

13 G. J. Wedemayer, L. H. Wang, P. A. Patten, P. G. Schultz, R. C. Stevens, *J. Mol. Biol.* **1997**, *268*, 390–400.

14 P. A. Patten, N. S. Gray, P. L. Yang, C. B. Marks, G. J. Wedemayer, J. J. Boniface, R. C. Stevens, P. G. Schultz, *Science* **1996**, *271*, 1086–1091.

15 G. J. Wedemayer, P. A. Patten, L. H. Wang, P. G. Schultz, R. C. Stevens, *Science* **1997**, *276*, 1665–1669.

16 J. D. Stewart, S. J. Benkovic, *Nature* **1995**, *375*, 388–391.

17 H. D. Ulrich, E. Mundorff, B. D. Santarsiero, E. M. Driggers, R. C. Stevens, P. G. Schultz, *Nature* **1997**, *389*, 271–275.

18 T. Asada, H. Gouda, P. A. Kollman, *J. Am. Chem. Soc.* **2002**, *124*, 12535–12542.

19 D. Wahler, J.-L. Reymond, *Curr. Opin. Chem. Biol.* **2001**, *5*, 152–158.

20 D. S. Tawfik, R. Chap, B. S. Green, M. Sela, Z. Eshhar, *Proc. Natl. Acad. Sci. U.S.A.* **1995**, *92*, 2145–2149.

21 W. P. Jencks, *Adv. Enzymol.* **1975**, *43*, 219–410.

22 J. Xu, Q. Deng, J. Chen, K. N. Houk, J. Bartek, D. Hilvert, I. A. Wilson, *Science* **1999**, *286*, 2345–2348.

23 D. Hilvert, K. W. Hill, K. D. Nared, M.-T. M. Auditor, *J. Am. Chem. Soc.* **1989**, *111*, 9261–9262.

24 D. C. Braisted, P. G. Schultz, *J. Am. Chem. Soc.* **1990**, *112*, 7430–7431.

25 V. E. Gouverneur, K. N. Houk, B. de Pascual-Teresa, B. Beno, K. D. Janda, R. A. Lerner, *Science* **1993**, *262*, 204–208.

26 A. A. P. Meekel, M. Resmini, U. K. Pandit, *J. Chem. Soc., Chem. Commun.* **1995**, 571–572; J. T. Yli-Kauhaluoma, J. A. Ashley, C.-H. Lo, L. Tucker, M. M. Wolfe, K. D. Janda, *J. Am. Chem. Soc.* **1995**, *117*, 7041–7047; M. Resmini, A. A. P. Meekel, U. K. Pandit, *Pure Appl. Chem.* **1996**, *68*, 2025–2028.

27 T. M. Tarasow, S. L. Tarasow, B. E. Eaton, *Nature* **1997**, *389*, 54–57; B. Seelig, A. Jäschke, *Chem. Biol.* **1999**, *6*, 167–176.

28 K. Auclair, A. Sutherland, J. Kennedy, D. J. Witter, J. P. Van den Heever, C. R. Hutchinson, J. C. Vederas, *J. Am. Chem. Soc.* **2000**, *122*, 11 519–11 520; K. Katayama, T. Kobayashi, H. Oikawa, M. Honma, A. Ichihara, *Biochim. Biophys. Acta* **1998**, *1384*, 387–395; K. Watanabe, T. Mie, A. Ichihara, H. Oikawa, M. Honma, *J. Biol. Chem.* **2000**, *275*, 38393–38401; T. Ose, K. Watanabe, T. Mie, M. Honma, H. Watanabe, M. Yao, H. Oikawa, I. Tanaka, *Nature* **2003**, *422*, 185–189.

29 J. Chen, Q. Deng, R. Wang, K. N. Houk, D. Hilvert, *ChemBioChem* **2000**, *1*, 255–261.

30 M. R. Haynes, M. Lenz, M. J. Taussig, I. A. Wilson, D. Hilvert, *Isr. J. Chem.* **1996**, *36*, 151–159.

31 F. E. Romesberg, B. Spiller, P. G. Schultz, R. C. Stevens, *Science* **1998**, *279*, 1929–1933.

32 F. E. Romesberg, P. G. Schultz, *Bioorg. Med. Chem. Lett.* **1999**, *9*, 1741–1744.

33 A. Piatesi, D. Hilvert, *ChemBioChem* **2004**, *5*, 460–466.

34 A. Piatesi, D. Hilvert, *Can. J. Chem.* **2002**, *80*, 657–664.

35 E. T. Boder, K. S. Midelfort, K. D. Wittrup, *Proc. Natl. Acad. Sci. U.S.A.* **2000**, *97*, 10701–10705; J. Hanes, L. Jermutus, S. Weber-Bornhauser, H. R. Bosshard, A. Plückthun, *Proc. Natl. Acad. Sci. U.S.A.* **1998**, *95*, 14130–14135.

36 M. L. Casey, D. S. Kemp, K. G. Paul, D. D. Cox, *J. Org. Chem.* **1973**, *38*, 2294–2301; D. S. Kemp, M. L. Casey, *J. Am. Chem. Soc.* **1973**, *95*, 6670–6680; D. S. Kemp, D. D. Cox, K. G. Paul, *J. Am. Chem. Soc.* **1975**, *97*, 7312–7318.

37 S. N. Thorn, R. G. Daniels, M.-T. M. Auditor, D. Hilvert, *Nature* **1995**, *373*, 228–230.

38 Y. Hu, K. N. Houk, K. Kikuchi, K. Hotta, D. Hilvert, *J. Am. Chem. Soc.* **2004**, *126*, 8197–8205.

39 F. P. Seebeck, D. Hilvert, *J. Am. Chem. Soc.* **2005**, *127*, 1307–1312.

40 A. J. Kirby, *Adv. Phys. Org. Chem.* **1980**, *17*, 183–278.

41 F. Hollfelder, A. J. Kirby, D. S. Tawfik, K. Kikuchi, D. Hilvert, *J. Am. Chem. Soc.* **2000**, *122*, 1022–1029.

42 E. Debler, S. Ito, F. P. Seebeck, A. Heine, D. Hilvert, I. A. Wilson, *Proc. Natl Acad. Sci. USA*, in press.

43 A. Genre-Grandpierre, C. Tellier, M.-J. Loirat, D. Blanchard, D. R. W. Hodgson, F. Hollfelder, A. J. Kirby, *Bioorg. Med. Chem. Lett.* **1997**, *7*, 2497–2502.

44 B. Golinelli-Pimpaneau, O. Goncalves, T. Dintinger, D. Blanchard, M. Knossow, C. Tellier, *Proc. Natl. Acad. Sci. U.S.A.* **2000**, *97*, 9892–9895.

45 S. L. Bearne, R. Wolfenden, *Biochemistry* **1997**, *36*, 1646–1656.

46 J. Na, K. N. Houk, D. Hilvert, *J. Am. Chem. Soc.* **1996**, *118*, 6462–6471.

47 H. Suga, O. Ersoy, S. F. Williams, T. Tsumuraya, M. N. Margolies, A. J. Sinskey, S. Masamune, *J. Am. Chem. Soc.* **1994**, *116*, 6025–6026.

48 T. Tsumuraya, H. Suga, S. Meguro, A. Tsunakawa, S. Masammune, *J. Am. Chem. Soc.* **1995**, *117*, 11390–11396.

49 P. Wirsching, J. A. Ashley, C.-H. L. Lo, K. D. Janda, R. A. Lerner, *Science* **1995**, *270*, 1775–1782.

50 F. Tanaka, H. Almer, R. A. Lerner, C. F. Barbas III, *Tetrahedron Lett.* **1999**, *40*, 8063–8066.

51 K. D. Janda, L.-C. Lo, C.-H. L. Lo, M.-M. Sim, R. Wang, C.-H. Wong, R. A. Lerner, *Science* **1997**, *275*, 945–948.

52 C. F. Barbas III, A. Heine, G. Zhong, T. Hoffmann, S. Gramatikova, R. Björnestedt, B. List, J. Anderson, E. A. Stura, I. A. Wilson, R. A. Lerner, *Science* **1997**, *278*, 2085–2092.

53 T. Hoffmann, G. Zhong, B. List, D. Shabat, J. Anderson, S. Gramatikova, R. A. Lerner, C. F. Barbas III, *J. Am. Chem. Soc.* **1998**, *120*, 2768–2779.

54 A. Heine, G. DeSantis, J. G. Luz, M. Mitchell, C. H. Wong, I. A. Wilson, *Science* **2001**, *294*, 369–374.

55 G. F. Zhong, T. Hoffmann, R. A. Lerner, S. Danishefsky, C. F. Barbas III, *J. Am. Chem. Soc.* **1997**, *119*, 8131–8132.

56 B. List, D. Shabat, G. Zhong, J. M. Turner, A. Li, T. Bui, J. Anderson, R. A. Lerner, C. F. Barbas III, *J. Am. Chem. Soc.* **1999**, *121*, 7283–7291; J. M. Turner, T. Bui, R. A. Lerner, C. F. Barbas III, B. List, *Chem. Eur. J.* **2000**, *6*, 2772–2774.

57 S. C. Sinha, C. F. Barbas III, R. A. Lerner, *Proc. Natl. Acad. Sci. U.S.A.* **1998**, *95*, 14603–14608.

58 B. List, D. Shabat, C. F. Barbas III, R. A. Lerner, *Chem. Eur. J.* **1998**, *4*, 881–885.

59 G. F. Zhong, R. A. Lerner, C. F. Barbas, *Angew. Chem. Int. Ed.* **1999**, *38*, 3738–3741.

60 P. G. Schultz, R. A. Lerner, *Acc. Chem. Res.* **1993**, *26*, 391–395.

61 B. F. Cravatt, J. A. Ashley, K. D. Janda, D. L. Boger, R. A. Lerner, *J. Am. Chem. Soc.* **1994**, *116*, 6013–6014.

62 K. D. Janda, C. G. Shevlin, R. A. Lerner, *Science* **1993**, *259*, 490–493.

63 A. Simeonov, M. Matsushita, E. A. Juban, E. H. Z. Thompson, T. Z. Hoffman, A. E. Beuscher, M. J. Taylor, P. Wirsching, W. Rettig, J. K. McCusker, R. C. Stevens, D. P. Millar, P. G. Schultz, R. A. Lerner, K. D. Janda, *Science* **2000**, *290*, 307–313.

64 H. Miyashita, Y. Karaki, M. Kikuchi, I. Fujii, *Proc. Natl. Acad. Sci. U.S.A.* **1993**, *90*, 5337–5340.

65 D. A. Campbell, B. Gong, L. M. Kochersperger, S. Yonkovich, M. A. Gallop, P. G. Schultz, *J. Am. Chem. Soc.* **1994**, *116*, 2165–2166.

66 D. Shabat, H. N. Lode, U. Pertl, R. A. Reisfeld, C. Rader, R. A. Lerner, C. F. Barbas, *Proc. Natl. Acad. Sci. U.S.A.* **2001**, *98*, 7528–7533.

67 B. Mets, G. Winger, C. Cabrera, S. Seo, S. Jamdar, G. Yang, K. Zhao, R. J. Briscoe, R. Almonte, J. H. Woods, D. W. Landry, *Proc. Natl. Acad. Sci. U.S.A.* **1998**, *95*, 10176–10181.

68 D. Hilvert, S. H. Carpenter, K. D. Nared, M.-T. M. Auditor, *Proc. Natl. Acad. Sci. U.S.A.* **1988**, *85*, 4953–4955.

69 Y. Tang, J. B. Hicks, D. Hilvert, *Proc. Natl. Acad. Sci. U.S.A.* **1991**, *88*, 8784–8786.

70 Y. Tang **1996**, *Evolutionary Studies with a Catalytic Antibody*, PhD thesis, The Scripps Research Institute.

71 G. MacBeath, D. Hilvert, *Chem. Biol.* **1996**, *3*, 433–445.

72 G. P. Smith, V. A. Petrenko, *Chem. Rev.* **1997**, *97*, 391–410.

73 G. Georgiou, C. Stathopoulos, P. S. Daugherty, A. R. Nayak, B. L. Iverson, R. Curtiss, *Nat. Biotechnol.* **1997**, *15*, 29–34; K. D. Wittrup, *Curr. Opin. Biotechnol.* **2001**, *12*, 395–399.

74 J. Hanes, A. Plückthun, *Proc. Natl. Acad. Sci. U.S.A.* **1997**, *94*, 4937–4942.

75 R. W. Roberts, J. W. Szostak, *Proc. Natl. Acad. Sci. U.S.A.* **1997**, *94*, 12297–12302.

76 S. V. Taylor, P. Kast, D. Hilvert, *Angew. Chem. Int. Ed.* **2001**, *40*, 3310–3335.

77 M. A. Dwyer, L. L. Looger, H. W. Hellinga, *Science* **2004**, *304*, 1967–1971.

5
Protein-based Artificial Enzymes

Ben Duckworth and Mark D. Distefano

5.1
Introduction

The design of catalysts that rival the specificity and speed of natural enzymes is a challenging objective. Various approaches have been employed to create such molecules. Polypeptides are a logical choice of possible framework for catalyst design because they provide a simple means for generating complex structures. The catalyst structure can be changed by altering the corresponding gene and expressing the desired protein in bacteria. However, chemical methods can also be incorporated into such an approach to greatly expand the range of possible reactions that can be studied. In such a chemogenetic strategy, a small molecule-based catalyst is coupled to a protein scaffold devoid of catalytic activity. The small molecule provides intrinsic reactivity while the protein component controls specificity; the protein can also be used to tune or augment the reactivity of the small molecule catalyst. Such an approach provides enormous flexibility in the design of new catalytic materials.

Early efforts in this field focused on using chemical modification to alter the type of reaction catalyzed by existing enzymes or to change their substrate specificity. Thiosubtilisin [1, 2], selenosubtilisin [3] and flavopapain [4] are all examples of proteins that manifest altered reactivity resulting from chemical modification. Other work with staphylococcal nuclease [5] and subtilisin [6, 7] employed chemical modification to modulate substrate selectivity. While that pioneering work served as the foundation for protein-based catalyst design, these efforts were, notably, based on protein frameworks that possessed native catalytic activity. Such proteins already contained highly evolved substrate binding sites and/or catalytic machinery. The present chapter focuses on the design of protein-based catalysts starting with polypeptides that have no intrinsic catalytic activity. While this is an enormous challenge when compared with approaches starting with existing enzymes, it opens up a plethora of exciting possibilities; if it could be accomplished, the ability to create catalysts with enzyme-like properties would have dramatic effects that would extend well beyond the field of chemistry. Several recent articles provide a more comprehensive review of the field of protein design, including work with existing enzymes [8–19].

5.2
Artificial Nucleases Based on DNA and RNA Binding Proteins

5.2.1
Introduction

The process of cleaving the phosphodiester bond of RNA and DNA is a vital and wide-spread phenomenon in nature. Scission of the nucleic acid backbone plays crucial roles in myriad in vivo and in vitro processes, including mutational repair, cell death, cloning, and sequencing. Much attention has been given to the discovery and development of nucleases, both natural and artificial, which might benefit the fields of biotechnology and pharmacology [20]. Several chemical agents possess nuclease activity. However, such compounds have low affinity for oligonucleotides and may not be optimally positioned for attack and scission of the backbone. Therefore, if a chemical nuclease could be incorporated into a DNA or RNA binding protein, substrate binding and nuclease activity would be enhanced.

5.2.2
Artificial Nucleases from Native Protein Scaffolds

To create an artificial nuclease, Sigman and co-worker enhanced the nuclease specificity of 1,10-phenanthroline(OP)-copper by covalently linking the moiety to the *Escherichia coli trp* repressor (TrpR) [21]. OP is a robust chemical nuclease that cleaves the phosphodiester bonds of nucleic acids. A copper-oxo species attacks the C-1H of the deoxyribose in the minor grove, yielding 3'- and 5'-phosphomonoesters, free base, and 5-methylene furanone (5-MF) (Figure 5.1) [22, 23]. This makes it possible to convert DNA-binding proteins into nucleases by conjugation with OP-copper. Since most DNA/RNA binding proteins bind the major groove, it is crucial for the OP-copper group to interact solely with the minor groove to avoid disruption of key binding interactions.

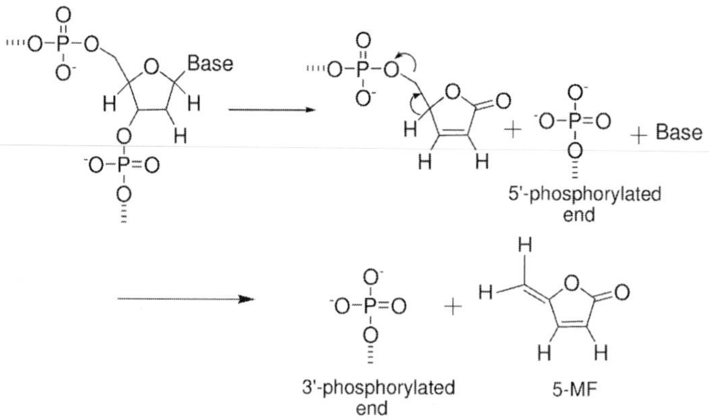

Figure 5.1 Shortened chemical mechanism for DNA cleavage by 1,10-phenanthroline-copper complexes.

a)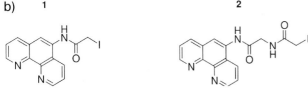

Figure 5.2 (A) Procedure for the attachment of 1,10-phenanthroline to proteins via lysine derivatization. (B) 5-(iodoacetamido)-1,10-phenanthroline (IOP, **1**) and 5-(iodoacetylglycylamido)-1,10-phenanthroline (IAOP, **2**).

To conjugate *E. coli trp* repressor with OP-copper, the four lysine amino groups were first converted into sulfhydryl groups by reaction with 2-iminothiolane hydrochloride in the presence of the corepressor L-tryptophan (Figure 5.2) [21, 24, 25]. The thiols were then alkylated with 5-(iodoacetamido)-1,10-phenanthroline (IOP, Figure 5.2-**1**), yielding the TrpR-OP conjugate. DNase I footprinting experiments indicated that the DNA-binding ability of TrpR was not jeopardized after modification with OP. Early experiments involved the incubation of *trp*EDCBA operator with OP-TrpR copper complex for 20 h, which resulted in double-strand scission as well as single-strand nicks in 50% of the labeled strands [21].

5.2.3
OP Nuclease Design by Mutagenesis and Chemical Modification

An additional approach to artificial nuclease design is cysteine mutagenesis followed by conjugation with the desired catalytic group. This avoids the need to convert amino acid residues into reactive thiols, while allowing precise control over the placement of the reagent within the DNA-binding protein. Sigman and co-workers applied this strategy to create an artificial nuclease using the bacteriophage λ Cro protein [26]. When complexed with DNA, the C-terminal arm binds within the minor groove and is close to the C-1 hydrogen of the deoxyribose on either DNA strand [26–30]. Therefore, an alanine close to the C-terminus was mutated to a cysteine and derivatized with IOP. After incubation with the Cro A66C-OP conjugate, 40% of the 17 base pair OR-3 operator site was cleaved within 10 min [11]. Thus, placement of the phenanthroline group close to the DNA substrate ensured efficient nuclease activity.

Recently, the carboxy terminal domain of NarL, NarL^C, was modified with 1,10-phenanthroline. NarL is a response regulatory protein of *E. coli* and binds to a heptameric consensus sequence [31]. Two residues of the C-terminal domain, which were mutated and conjugated with IOP (NarL^C K201C-IOP, NarL^C K211C-IOP), showed high site-specific cleavage efficiency of the top strand (Figure 5.3). When the mutated NarL^C was modified with IAOP (Figure 5.2-2), a similar DNA cleavage pattern was observed on the bottom strand. This IAOP conjugate contains a 4 Å longer linker arm than does the

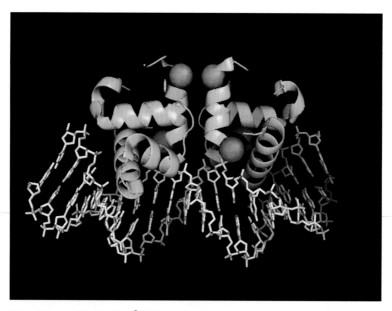

Figure 5.3 Model of the NarL^C-DNA complex showing the locations of amino acids 201 and 211. Two molecules of NarL^C are positioned above the major groove of the DNA. Color scheme: Protein secondary structure (green), DNA (white), residue 211 (pink), residue 201 (orange).

previous IOP conjugate, which allows the OP-Cu complex to extend and cleave the adjacent bases of the DNA substrate. These artificial cleavage enzymes could ultimately be used to identify the number, position and orientation of NarL monomers on several promoters and may be extended to other regulatory proteins. These results highlight the promise of chemical nucleases as tools for biological investigation.

5.2.4
Additional Applications for OP Conjugates

Other research groups have also created artificial nucleases by OP conjugation. Johnson and co-workers created an artificial nuclease, which aided in the understanding of DNA conformational changes upon protein binding [32]. *E. coli* Fis (factor for inversion stimulation) was chosen as the protein scaffold and binds to DNA with a low sequence specificity. X-Ray crystallography and electrophoretic mobility studies indicated significant bending of the DNA upon Fis binding [33, 34]. To aid in the study of DNA-Fis complexes, OP-Cu was conjugated to four separate mutated cysteines. Of these four conjugates, N98C-OP and N78C-OP displayed nuclease activity. Scission patterns from these two nucleases not only confirmed earlier studies of DNA bending, but also revealed that the DNA wrapped around the Fis dimer.

The *E. coli* catabolite gene activator protein (CAP) binds to a 22 base-pair, two-fold symmetric DNA recognition site with high affinity ($K_A = 4 \times 10^{10}$ M^{-1}) [35]. Ebright and co-workers successfully converted CAP into a site-specific DNA cleavage enzyme. In their first attempt, OP was conjugated to the only solvent accessible cysteine (C_{178}) of CAP. The catalytic group was placed at this position so that intra-CAP interactions would not be disrupted, CAP-DNA contacts would not be comprised, and OP would be in close proximity for a favorable attack on the phosphodiester bond. The binding constant for this conjugate was 1×10^8 M^{-1}, indicating that the DNA binding ability of CAP was not significantly lost due to conjugation with OP. The CAP178-OP conjugate cleaved the DNA site at four adjacent nucleotides (Figure 5.4).

Earlier work indicated that in the specific CAP-DNA complex the DNA was bent 90° away from the protein but was not distorted in the nonspecific complex. (CAP bound to its noncognate DNA site) [3, 36]. This DNA bending phenomenon was exploited to construct a new CAP-OP nuclease capable of single site cleavage [37]. The crystal structure of the specific CAP-DNA complex revealed that amino acids 24–26 and 89–91 of CAP were close to the DNA substrate [38]. With this in mind, residue 26 was mutated

Figure 5.4 DNA cleavage sites produced by the CAP178–1,10-phenanthroline conjugate. Longer arrows indicate sites where greater cleavage efficiency occurred.

to a cysteine and modified with IOP. While the catalytic group neighbors the DNA in the specific complex, OP is too far from the substrate in the nonspecific complex, and thus cannot cleave the DNA. Ultimately, conjugation of OP at position 26 yielded an artificial nuclease that proceeded to ~90% cleavage with no detectable nonspecific cleavage at distal sites.

5.2.5
A Fe-EDTA Artificial Nuclease

An alternative DNA-cleaving agent was used to identify the σ subunit DNA contact sites of *E. coli* RNA polymerase [39]. Iron-EDTA protein conjugates have previously been shown to cleave DNA by producing hydroxyl radicals, which ultimately attack the deoxyribose backbone [40]. To create this artificial nuclease, Minchin and co-workers mutated several residues within the σ subunit to cysteines. These residues were conjugated with (S)-1-[p-(bromoacetamido)benzyl]-EDTA (BABE) (Figure 5.5). By identifying the cleavage products of several related promoters, this Fe-BABE conjugate confirmed the location of the σ subunit-promoter DNA contact sites.

5.2.6
Concluding Remarks

Through direct conjugation of chemical nucleases to DNA binding proteins, several groups have successfully created artificial nuclease from proteins with no native nuclease ability. These nucleases can utilize the favorable protein–DNA binding elements and can position the hydrolytic metal close to the DNA backbone for efficient cleavage [41]. These nucleases have provided useful biological insights and a better understanding of the molecular details involved in protein–DNA interactions.

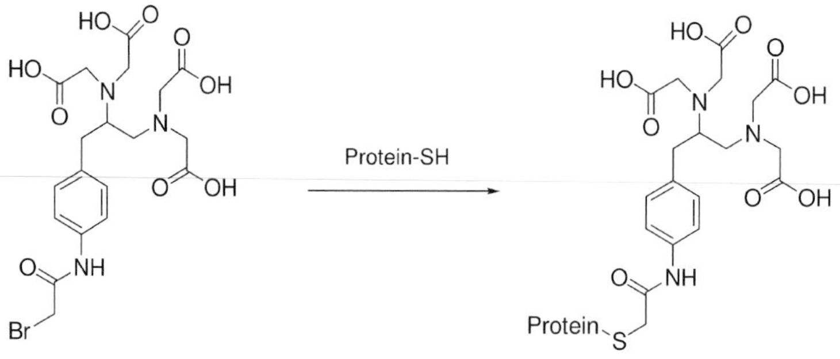

Figure 5.5 Procedure for the derivatization of proteins with (S)-1-[p-(bromoacetamido)benzyl]-EDTA (BABE).

5.3
Catalysts Based on Hollow Lipid-binding Proteins

5.3.1
Lipid-binding Proteins

Lipid-binding proteins are a class of molecules found in eukaryotic cells involved in the transport of fatty acids and other types of hydrophobic compounds. The protein structure consists of two orthogonal planes of β-sheets that form a cup-shaped cavity that is capped off with a helix-turn-helix element. Considerable structural data, obtained from X-ray and NMR analysis, exists for this family of proteins [42, 43]. Of particular interest is the presence of a large, solvent sequestered, cavity whose overall volume varies between 500 and 1000 A^3 depending on the exact identity of the protein; with the exception of certain membrane-bound channel proteins, there are few other examples of macromolecules that possess such a cavity. This structural feature serves as the ligand binding site in the protein cavity for a diverse range of molecules. Such a large, solvent sequestered, binding site also provides a useful scaffold for the design of catalysts. Formally, this cavity can be viewed as a protein equivalent to the cyclodextrin template used in much of the pioneering work of Breslow and co-workers in their development of enzyme mimics [18].

5.3.2
Initial Work

In their initial work Distefano and co-workers used the protein ALBP (adipocyte lipid-binding protein) for their protein scaffold. This protein contains a unique cysteine residue a position 117 that can be selectively modified using reagents that capitalize on the unique reactivity of the thiol side chain. Kuang and co-workers developed a reagent, TP-PX (**5.6-1**) that contained an activated disulfide suitable for protein derivatization. This molecule was used to incorporate a PX moiety into ALBP at position 117, resulting in the formation of a construct denoted ALBP-PX (Figure 5.6, **5.6-2**) [44]. This semisynthetic biocatalyst aminated reductively various α-keto acids (**5.7-2**) to amino acids (**5.7-4**) with 0 to 94% ee (Figure 5.7). Because these reactions were performed in the absence of any additional amine source, only single turnovers were obtained. The reaction rates were not, however, significantly faster than those involving free pyridoxamine (**5.7-1b**). This suggested that the protein cavity functions as a chiral environment that controls the facial selectivity of the protonation of the aldimine intermediate without forming specific interactions with the bound pyridoxamine cofactor, which could accelerate the reaction, as confirmed by a X-ray crystal structure [45]. Modeling of the Schiff base complexes with several amino acids indicated that one face of the putative aldimine intermediate was protected against the approach of the solvent or buffer molecules that must be the proton source for the reaction given the lack of suitable functional groups within the cavity. This structural data provided a rationale for explaining the enantioselectivity observed in the ALBP-PX system.

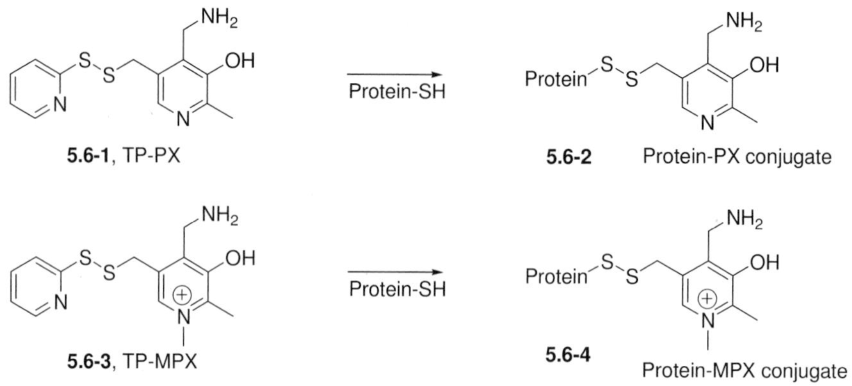

Figure 5.6 Derivatization reagents for the preparation of pyridoxamine conjugates of fatty acid binding proteins.

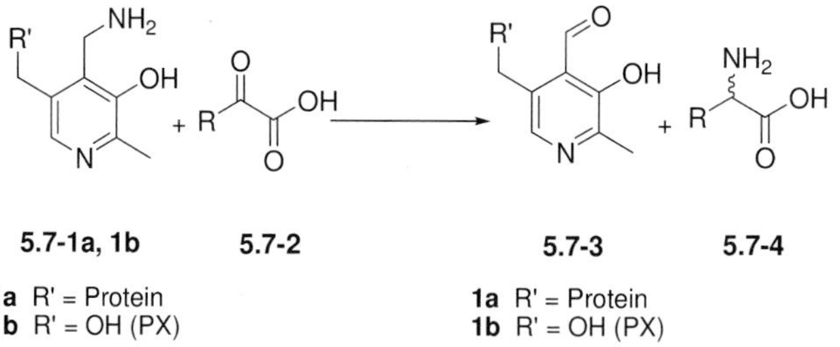

5.7-1a, 1b   **5.7-2**   **5.7-3**   **5.7-4**

1a R' = Protein
1b R' = OH (PX)

1a R' = Protein
1b R' = OH (PX)

Figure 5.7 Single turnover, reductive amination, reactions promoted by fatty acid binding protein–pyridoxamine conjugates.

5.3.3
Exploiting the Advantage of a Protein-based Scaffold

The next step in developing these protein-based catalysts was to take advantage of the synthetic flexibility provided by a protein scaffold. Since the protein is encoded by a DNA sequence, changes can be made in the gene that will result in modifications in the protein. Mutagenesis and other genetic methods can produce alterations in the protein framework much more rapidly than the organic synthesis required to make changes in synthetic scaffolds. Initially, the position of catalyst attachment within the cavity was varied to probe its effect on reaction rate and selectivity. For these experiments, a different fatty acid binding protein (IFABP) was employed. The wild-type IFABP contains no cysteine residues in its primary sequence and is, hence, a useful template for the introduction of single cysteine residues at different positions [46]. Cysteine residues were introduced at several positions, including V_{60}, L_{72}, and A_{104}, using site-directed mutagenesis. Figure 5.8 gives the locations of these mutations within the protein structure.

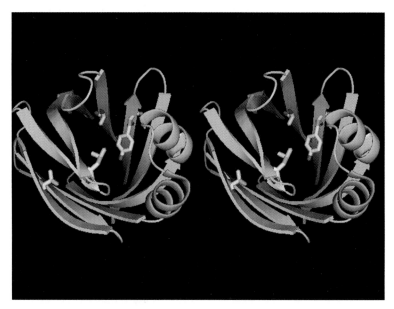

Figure 5.8 Stereo view showing positions 60, 72, 104 and 117 in IFABP that have been used for the attachment of pyridoxamine and other catalytic groups. From top to bottom: A104, Y117, L72 and V60. Color scheme: Protein secondary structure (green), carbon (white), oxygen (red), nitrogen (blue).

The three different mutant proteins described above were purified and the corresponding conjugates were prepared using the TP-PX reagent (**5.6-1**). The conjugates were then evaluated in single turnover conditions (Figure 5.7) with several α-keto acids to produce amino acids in enantiomerically enriched form. This collection of catalysts exhibited various differences in reactivity and selectivity. Compared to ALBP-PX, IFABP-PX60 reacted at least 9.4-fold more rapidly, while IFABP-PX72 displayed opposite enantioselectivity, and IFABP-PX104 showed a clear selectivity preference for unbranched substrates [47]. From these experiments, the position of cofactor attachment is clearly an important parameter in modulating the reactivity and specificity of these biocatalysts. Moreover, these results underscore the utility of site-directed mutagenesis and the power of using a protein-based scaffold for catalyst development.

5.3.4
Catalytic Turnover with Rate Acceleration

Due to its intriguingly more rapid reaction rate than ALBP-PX and free pyridoxamine, IFABP-PX60 was subsequently studied further [48]. Under single turnover conditions, it converted α-keto glutarate (Figure 5.9, **5.9-1**) into glutamic acid (**5.7-4**) 62-fold faster than free pyridoxamine (**5.7-1b**); this was determined by evaluating the extent of conversion at much shorter reaction times. This more rapid reaction rate under single

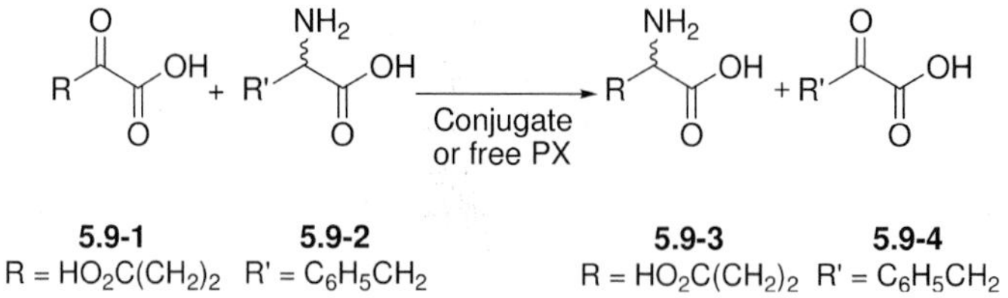

5.9-1   **5.9-2**   **5.9-3**   **5.9-4**
$R = HO_2C(CH_2)_2$ $R' = C_6H_5CH_2$ $R = HO_2C(CH_2)_2$ $R' = C_6H_5CH_2$

Figure 5.9 Multiple turnover reactions that produce chiral amino acids from achiral keto acids.

turnover conditions suggested that this construct might be capable of catalytic transamination. Using tyrosine or phenylalanine (**5.9-2**) as the amine source to recycle the cofactor from the pyridoxal form (**5.7-3a**) back to the pyridoxamine form (**5.7-1**), L-glutamate (**5.9-3**) was formed with an enantiomeric purity of 93% ee. As many as 50 turnovers were observed with long reaction times (14 days). Both k_{cat} and K_M for the reaction were determined by kinetic analysis (Table 5.1).

Comparison with similar parameters obtained from reactions with free pyridoxamine indicated that IFABP-PX60 catalyzed transamination some 200 times more efficiently. Analysis of the specific kinetic constants k_{cat} and K_M indicated that the observed rate acceleration was due mostly to an increase in substrate binding (50-fold), with a smaller effect on the maximal rate (4-fold). While this is an impressive result, the absolute magnitude of k_{cat}/K_M (0.02 s^{-1} M^{-1}) makes it clear that this catalyst is still quite primitive compared to natural enzyme systems that occasionally operate with catalytic efficiencies near the diffusion limit.

Table 5.1 Kinetic constants and catalytic efficiencies for semi-synthetic transaminases based on fatty acid binding proteins.

	K_M (mM)	k_{cat} (h^{-1})	k_{cat}/K_M (h^{-1} mM^{-1})
PX	73	0.032	4.4×10^{-4}
MPX	38.7	0.031	8.0×10^{-4}
IFABP-PX60	1.8	0.29	0.16
IFABP-MPX60	6.8	0.23	0.034
hsIFABP-PX60	10.2	0.22	0.022
IFABP-PxK38	0.81	0.44	0.54
IFABP-MPxK38	13.7	1.12	0.08
IFABP-PxK51	0.24	0.44	1.83
IFABP-MPxK51	8.9	0.52	0.06
IFABP-Px126	5.5	0.18	0.03
IFABP-MPx126	40	1.10	0.03
IFABP-Px126/14	6.0	0.11	0.02
IFABP-MPx126/14	48	0.78	0.02

5.3.5

Modulation of Cofactor Reactivity with Metal Ions

Model studies with pyridoxamine and related analogs have shown that transamination reactions promoted by these catalysts can be accelerated by the addition of metal ions, including Zn(II), Cu(II), Ni(II) and Al(III). Hypothetically, these cations accomplish this by stabilizing the formation of Schiff base intermediates and by increasing the acidity of the protons that must be removed in the reaction. Interestingly, the addition of metal ions to reactions catalyzed by ALBP- and IFABP-pyridoxamine conjugates resulted in both positive and negative reaction rate perturbations [49]. IFABP-PX104 reacted 4.7-fold faster in the presence of Cu(II) whereas IFABP-PX60 reacted 4.4-fold slower – both rate effects were accompanied by a decrease in reaction enantioselectivity. Little change was observed for the reaction catalyzed by IFABP-PX72. UV/Vis spectroscopy experiments that monitored the formation of metal-aldimine intermediates suggested that IFABP-PX60 and IFABP-PX104 but not IFABP-PX72 formed a complex with Cu(II). Thus, metal ions may be used to increase semisynthetic enzyme efficiency, although this did not occur in all cases. These studies also noted that the reactions rates were sensitive to changes in buffer; the use of imidazole resulted in a lower rate than with similar reactions performed in HEPES buffer. These observations raise the possibility that buffer molecules may actually enter the protein cavity and directly participate in the reaction. This makes sense because several proton transfers must occur in a complete transamination reaction and yet there are no functional groups within the cavity that could do this.

5.3.6

Chemogenetic Approach

In the catalytic mechanism of pyridoxamine-based reactions, the pyridine nitrogen atom must be protonated; this protonation gives the pyridine ring a net positive charge that increases the acidity of the nearby benzylic protons and serves to stabilize a developing negative charge. One way protonation can be effectively enforced is by introducing a permanent positive charge via N-methylation. In some model systems, an N-methylated species accelerates the transamination rate up to 20-fold. In contrast, after reconstitution with *N*-methylpyridoxamine, alanine aminotransferase lost its activity almost completely (>99.8%). Therefore, N-quaternization may have both positive and negative influences on the reaction rate. To allow a similar modification to be made in these fatty acid binding protein systems, a new reagent containing *N*-methylpyridoxamine and an activated disulfide (TP-MPX, **5.6-3**) was prepared. A conjugate, IFABP-MPX60 (**5.6-4**) was prepared [50] and, under catalytic conditions, significant reaction (5.2 turnovers, 41% ee) was observed using α-keto glutarate and phenylalanine as substrates. Kinetic analysis of the reaction showed that the MPX-containing conjugate gave higher K_M (3.8-fold higher than the earlier PX construct) but similar k_{cat}. This indicated that N-methylation had no positive effects on the reaction catalyzed by conjugates based on IFABP-V$_{60}$C. However, the MPX cofactor proved to be considerably more useful in subsequent experiments with IFABP variants containing lysine mutations (see below for further discussion of those proteins). In those studies, efforts were made to increase their catalytic efficiency by employing a genetic method to enforce N-protonation. Taking a cue from the crystal structure of AATase (aspartate amino transferase), an enzyme that catalyzes transamination using a pyridoxamine cofactor, carboxylate-containing amino acid residues were introduced close to the pyridine nitrogen locus in two IFABP mutant proteins. It was thought, based on functional studies of AATase, that the positioning of anionic side chains near the pyridine nitrogen center would enforce protonation via an ion pairing mechanism; the presence of the proximal negative charge would favor the protonated pyridine. Several carboxylate-containing IFABP mutants were prepared for this purpose. Unfortunately, none of these mutants were sufficiently stable for conjugate preparation – they precipitated during purification, and efforts to refold the denatured protein were unsuccessful. However, in separate experiments, the TP-MPX reagent (**5.6-3**) was used to prepare MPX-containing constructs instead (see **5.6-4** for a generic structure). Interestingly, these conjugates showed enhanced catalytic activity compared to their PX progenitors [51]. Kinetic studies indicated that the k_{cat} (1.12 h^{-1} for IFABP-MPxK38 and 0.52 h^{-1} for IFABP-MPxK51) and turnover numbers (12.2 turnovers by IFABP-MPxK38 and 5.7 by IFABP-MPx51 in 24 h) observed with these constructs under standard conditions are the highest achieved in this system (see Table 5.1 for a summary of kinetic parameters). The success of these constructs, prepared using a combination of chemical modification of the catalyst structure and genetic manipulation of the protein scaffold, highlights the enormous power and flexibility of this chemogenetic approach for catalyst development.

5.3.7
Adding Functional Groups within the Cavity

A major goal of research involving these protein scaffolds was to determine whether the flexibility of using a protein scaffold could be fully capitalized upon by using rational design in concert with knowledge of chemical mechanism to improve catalytic efficiency. Taking a cue from Nature, based on biochemical experiments with AATase, lysine residues were introduced into the protein cavity to enable Schiff base formation and to serve as general acids and bases in the reaction cycle. Figure 5.10 shows these functions in an abbreviated transamination mechanism.

Based on the crystal structure of IFABP, molecular modeling was employed to identify possible positions where lysine residues could be introduced to perform the functions noted above. Mutants $L_{38}K,V_{60}C$ and $E_{51}K,V_{60}C$ were prepared and used to create pyridoxamine conjugates [52]. Figure 5.11 gives a model of IFABP-PXK51. The resulting assemblies IFABP-PxK38 and IFABP-PxK51 showed improved k_{cat} and K_M. The overall catalytic efficiency (k_{cat}/K_M) of IFABP-PxK51 increased 4200-fold compared to unliganded pyridoxamine phosphate and was 12-fold greater than for IFABP-PX60 while maintaining comparable enantioselectivity (83–94% ee). The principal effect on the kinetic constants for the reactions catalyzed by these mutants was on K_M. Each conjugate showed a significant decrease in K_M (2.2- and 7.5-fold) and a small increase in k_{cat} (1.5-fold). UV/Vis spectroscopy, fluorescence and electron-spray mass

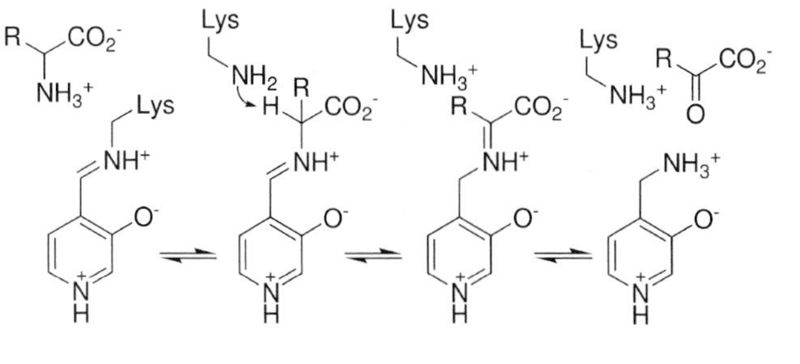

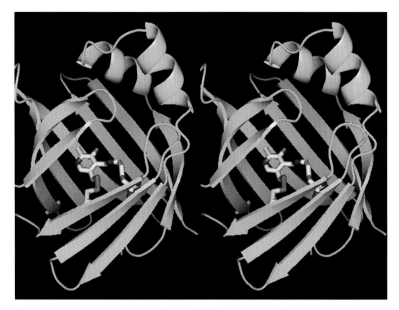

5.10-1
Pyridoxal
(in Schiff base)

5.10-2
Aldimine

5.10-3
Ketimine

5.10-4
Pyridoxamine

Figure 5.10 Abbreviated mechanism for transamination reactions promoted by pyridoxal-based catalysts.

spectrometry verified the catalytic function (Schiff base formation) of the introduced lysine residue in the reaction process. Significantly, in a study of rate versus pH, IFABP-PX60 gave a rate that decreased monotonically with increasing pH while the lysine mutants exhibited a bell-shaped profile with a maximum rate near pH 7.5. Taken together, these results provide compelling evidence that the lysines participate directly in the reaction and that features of enzymatic processes, including covalent catalysis, can be mimicked successfully.

Figure 5.11 Stereo representation of a model of IFABP-PXK51. The overall protein structure is shown with C_{60} and the pyridoxal linked via a disulfide bond. The pyridoxal aldehyde group is bonded through a Schiff base to K_{51}. Color scheme: Protein secondary structure (green), carbon (white), oxygen (red), nitrogen (blue), sulfur (orange).

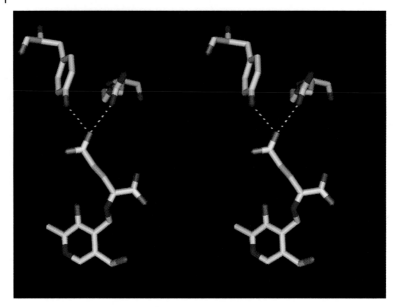

Figure 5.12 Putative binding site for the side-chain carboxylate of the substrate α-keto glutarate in IFABP-PX60. Present are Y_{14} (upper left), R_{126} (upper right) and the Schiff base complex between α-keto glutarate and the pyridoxamine catalyst. (lower center). Dashed lines indicate putative hydrogen bonds between the γ-carboxylate of α-keto glutarate and the side-chain phenol of Y_{14} and the guanidinium group from R_{126}. Color scheme: Carbon (white), oxygen (red), nitrogen (blue), sulfur (orange).

Computer modeling together with mutagenesis was also used to identify existing residues in the protein cavity that contributed to the substrate specificity of the IFABP-PX60 catalyst. R_{126} and Y_{14} were identified as two important residues that interacted with the γ-carboxylate group of α-keto glutarate when bound to the enzyme via a Schiff base with the PX moiety (Figure 5.12). Mutants IFABP-$V_{60}C,R_{126}M$ and -$V_{60}C,R_{126}M,Y_{14}F$ were prepared and pyridoxamine was attached to each mutant [53]. Of particular note, IFABP-PxM126 had a K_M three-fold higher than IFABP-PX, while the second mutation at position 14 had no significant effect. The k_{cat}s for both conjugates were 2-fold lower than for the original IFABP-PX60. N-methylated pyridoxamine conjugates were also introduced into these two mutants. Compared to the original IFABP-MPX60 conjugate described above, the mutations in positions 126 and 14 increased K_M as much as seven-fold. IFABP-MPxM126 gave a particularly significant result, manifesting a 5-fold higher k_{cat}. While neither the experiments with the lysine mutants nor those with the carboxylate binding site mutants produced catalysts that rival the efficiency of natural enzymes, incremental improvements were observed with many of the conjugates. Moreover, these advances came from constructs that combined mutations in the protein scaffold with synthetic alterations in cofactor structure, thus highlighting the power of the chemogenetic approach.

5.3.8
Scaffold Redesign

While site-directed mutagenesis for the incorporation of new functional groups into the protein cavity is the most obvious type of genetic modification that can be performed with these systems, more global changes can be made in the scaffold using similar methods. For example, as noted above, the IFABP architecture consists of a β-barrel structure caped off with a α-helical lid. To examine whether entry or exit of substrates or products into or out of this closed cavity was limiting catalytic turnover, a helixless (hs) IFABP mutant was prepared by deleting 17 helical residues (15–31) from the N-terminus and replacing them with a dipeptide linker (Ser-Gly) using site-directed mutagenesis [54]. A previous NMR study showed that this mutant preserved the original β-sheet secondary structure without the α-helical lid and was still relatively stable [55]. The structures of IFABP and hsIFABP are compared in Figure 5.13; except for the absence of the α-helical lid, the overall protein fold is largely intact. Studies that investigated reactions with this conjugate (hsIFABP-PX60) revealed that in 24 h, approximately two turnovers were observed with a selectivity of 93 % ee. A kinetics analysis indicated that k_{cat}s were comparable for both IFABP-PX60 (0.20 h^{-1}) and

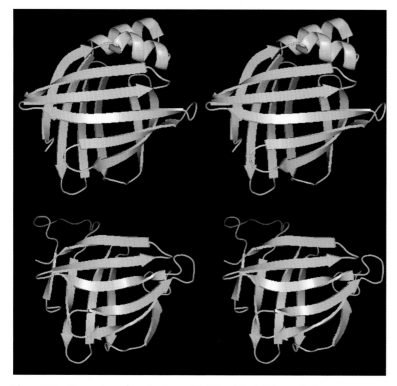

Figure 5.13 Comparison of the structures of IFABP and the helixless variant hsIFABP. Top: Stereo view of IFABP. Bottom: Stereo view of hsIFABP.

hsIFABP-PX60 (0.22 h^{-1}), while the former exhibited a 4-fold increase in K_M [56]. After introducing the MPX cofactor into the cavity of helixless protein, 4.9 turnovers in 24 h with 35 % ee was observed for the production of Glu under the same conditions. Taken together, these results suggested that removal of the α-helical lid did not affect the maximal rate and enantioselectivity of the respective constructs. However, deletion of this structural element did decrease the substrate binding affinity, as evidenced by an increase in K_M. Thus, it appears that while the hsIFABP scaffold was not useful in increasing the efficiency of PX-promoted transamination, these experiments did show that stable constructs could be prepared based on the helixless scaffold. This may have greater utility in reactions with larger cofactor catalysts or substrates.

5.3.9
Hydrolytic Reactions

Fatty acid binding proteins have also been used as scaffolds for constructing hydrolytic catalysts. Phenanthroline was attached to Cys$_{117}$ of ALBP using iodoacetamido 1,10-phenanthroline and metallated to produce a coordinated Cu(II) ion encapsulated within a chiral protein cavity (Figure 5.14). The resulting conjugate, ALBP-Phen-Cu(II) was able to promote the hydrolysis of several unactivated amino acid esters under mild conditions (pH 6.1, 25 °C) at rates 32- to 280-fold above the background rate in buffered aqueous solution; these reactions also showed modest stereoselectivity [57]. In 24 h incubations, 0.70–7.6 turnovers were obtained with enantiomeric excesses ranging from 31 to 86 % ee. ALBP-Phen-Cu(II) could also catalyze the hydrolysis of an activated amide (picolinic acid methyl nitroanilide, PMNA), under slightly more vigorous conditions (37 °C) and after longer incubation times. The rate of amide hydrolysis was 1.6 × 10^4-fold higher than the background rate [57]. However, k_{cat} obtained with ALBP-Phen-Cu(II) was still significantly lower than that obtained with a related Cu(II) bipyridine complex [58]. This rate decrease may reflect a non-optimal, perhaps nonplanar, conformation for PMNA binding within the ALBP cavity. Of related interest, an X-ray crystal structure of ALBP-Phen was obtained. Inspection of this structure showed that the protein could not accommodate the phenanthroline and PMNA within a planar conformation without significant distortion of the protein backbone [45]. This may account for the lower than expected rate of PMNA hydrolysis promoted by ALBP-Phen-Cu(II) compared with non-proteinaceous model complexes.

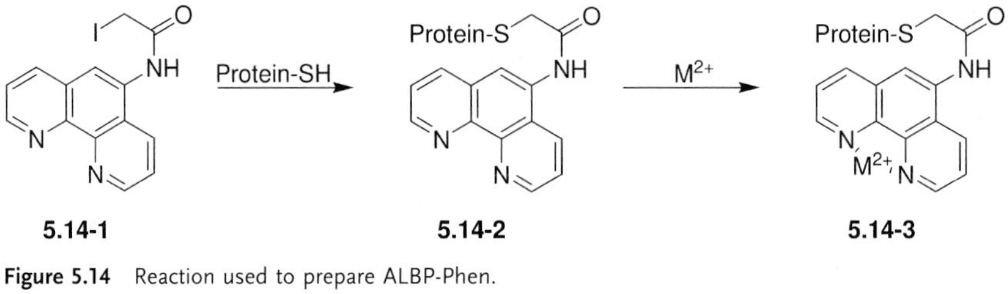

5.14-1　　　　　　　　　　**5.14-2**　　　　　　　　　　**5.14-3**

Figure 5.14　Reaction used to prepare ALBP-Phen.

In analogy to the above experiments with pyridoxamine complexes, the 1,10-phenan-throline ligand was introduced at several alternative sites (positions 60, 72 and 104, see Figure 5.8) using the protein scaffold IFABP [59]. Using alanine isopropyl ester as a substrate, IFABP-Phen60 catalyzed ester hydrolysis with less selectivity than ALBP-Phen, while Phen72 promoted the same reaction with higher selectivity. In contrast, hydrolysis of tyrosine methyl ester was catalyzed with higher selectivity by Phen60 and more rapidly by Phen104. These results indicated that the rate enhancement and substrate selectivity of hydrolysis reactions catalyzed by phenanthroline conjugates depended largely on the orientation and environment of the metal ligand within the protein cavity.

5.3.10
A Flavin-containing Conjugate

Reactions catalyzed by a flavin analog incorporated into IFABP have also been studied by alkylating a Cys residue within the cavity of the helixless variant (hsIFABP) [60]. The conjugate hsIFABP-FL catalyzed the oxidation of several dihydronicotinamides. These experiments were performed mainly to compare results obtained with flavin-IFABP conjugates with similar results acquired with flavopapains; these latter proteins were among the first semisynthetic enzymes produced. Interestingly, while hsIFABP-FL and flavopapain gave comparable rate accelerations (k_{cat}/K_M) for dihydronicotinamide oxidation, hsIFABP-FL manifested much higher K_M and k_{cat} than did flavopapain. The low K_M observed with flavopapain indicates that it primarily accelerates the reaction rate by enhancing substrate binding, whereas the higher k_{cat} obtained with hsIFABP-FL suggests that its major mode of rate acceleration involves enhancing flavin reactivity. Molecular modeling of hsIFABP-FL indicates that Gln_{27} is close to N(3)H and O(4) of the flavin (Figure 5.15). Interaction between the carboxamide of Gln_{27} and the flavin is likely to alter the redox potential of the isoalloxazine and hence augment its reactivity. Taken together, these results with flavopapain and hsIFABP-FL highlight the diverse ways in which protein scaffolds can be used to modulate catalyst activity.

5.3.11
Some Limitations

While the examples described above highlight the impressive results that can be achieved with protein as scaffolds for catalyst design, this approach is not without problems. Firstly, there is the issue of cavity size. It would be useful to incorporate large metal–ligand systems into this protein system to generate catalysts for a plethora of

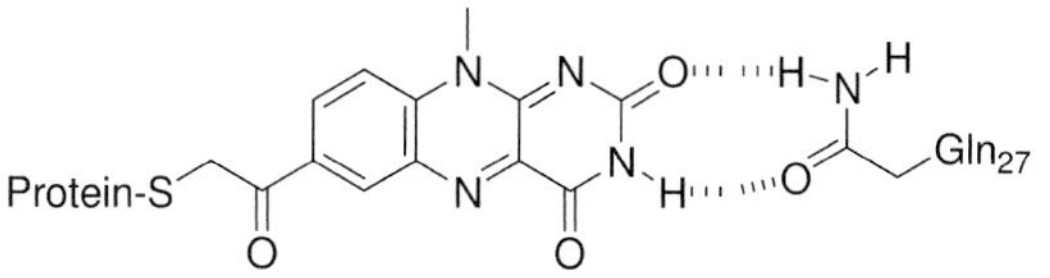

Figure 5.15 Hydrogen bonding to the flavin in hsIFABP-FL.

interesting reactions. Unfortunately, the enclosed cavity of fatty acid binding proteins (and even the somewhat open cavity of hsIFABP) limits the size of what can be introduced into the protein interior. Efforts to incorporate a heme group or a Mn(III)-salen complex into hsIFABP were not successful. While it was possible to attach these moieties to the protein scaffold under denaturing conditions and refold them, the resulting conjugates were unstable and underwent aggregation and precipitation upon storage. In future work, it will be necessary to identify other protein structures that can serve as effective scaffolds but that are not subject to these size limitations. Secondly, work with fatty acid binding proteins is limited to aqueous systems. This creates two types of problems. The first is that catalytic systems that are sensitive to water cannot be employed. The Mn(III)-salen complex described above underwent Schiff base hydrolysis and subsequent ligand decomposition upon prolonged storage in aqueous buffer. While the rate of this hydrolytic degradation can be decreased by employing mixtures of water and organic solvents, fatty acid binding proteins have limited solubility in such solvents. In addition, the inability to work in such solvents also limits the substrate concentrations that can be used and also precludes the use of these catalysts with many types of hydrophobic substrates. However, given the extensive interest and the large body of literature concerning the use of enzymes in organic media, it should be possible to identify other types of proteins that can serve as scaffolds for catalyst design in such solvent systems.

5.4
Myoglobin as a Starting Point for Oxidase Design

5.4.1
Artificial Metalloproteins and Myoglobin

Metalloproteins represent a major faction of scaffold proteins in artificial enzyme design. The introduction of a wide range of metals and redox cofactors into protein scaffolds can greatly increase both the diversity of enzyme catalysis and the application of these artificial metalloproteins. To sequester artificial redox cofactors into protein scaffolds, which would otherwise not bind these chemical catalysts, one may either use non-covalent or covalent attachment [61]. Myoglobin is a well-studied protein that contains a non-covalently bound heme for oxygen transport. The large heme-binding site provides substantial room to accommodate other ligand systems, and for these reasons serves as a useful starting point for catalyst design.

5.4.2
Non-covalent Attachment of a Redox Center

Watanabe and co-workers pursued an approach involving the non-covalent placement of Mn (III) and Cr (III) salophen complexes into apo-myoglobin [61]. In this artificial metalloprotein, two residues required mutation to improve the binding affinity of the cofactor and to allow increased access of substrates. The sulfoxidation activity of this Cr(III)-salophen myoglobin complex was studied and showed a six-fold rate increase over free Cr(III)-salophen in solution. The enantioselectivity of free salophen was 0%, while that of the myoglobin catalyst was 13%. While the ee is low for this artificial metalloprotein, Wantabe and co-workers were successful in demonstrating that asymmetric reactions can be accomplished using chiral protein cavities.

5.4.3
Dual Anchoring Strategy

To increase the enantioselectivity of these myoglobin metalloenzymes, Lu and co-workers have successfully utilized a covalent linkage approach [62]. In an earlier attempt a Mn(III)-salen complex was incorporated into apo-myoglobin by mutating residue 103 to cysteine, followed by modification with a methane thiosulfonate derivative of Mn(III)(salen) (Figure 5.16). This catalyst showed sulfoxidation activity; however, the ee was only 12%. As such a low ee might be a result of the ability of the bound ligand to exist in multiple conformations within the protein cavity, it was hypothesized that the rotational freedom of the salen complex could be limited if it was anchored at

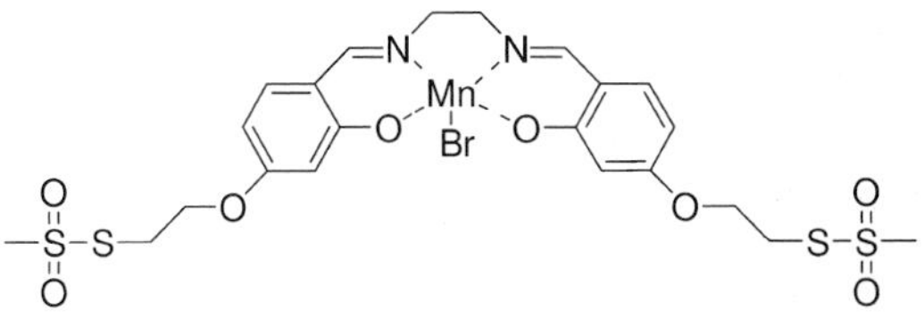

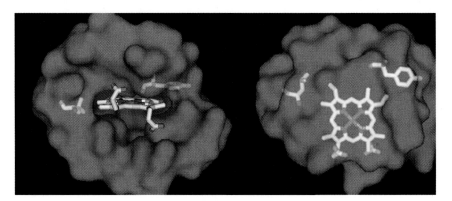

Figure 5.16 A Mn(III)(salen) derivatizing reagent used to prepare protein-Mn(III)(salen) conjugates.

Figure 5.17 Structure of myoglobin complexed with a heme group. The two residues chosen for mutation, Y_{103} and L_{72}, are shown. Left: Side view of myoglobin. Right: Top view of myoglobin. Color scheme: Carbon (white), oxygen (red), nitrogen (blue), iron (pink).

two separate locations, unlike the previous single-point attachment conjugate. The Mn(III)(salen) derivative was modeled into myoglobin to identify two positions that would yield the best fit into the heme pocket (Figure 5.17). In addition to an increase in the rate of sulfoxidation of the substrate thioanisole, the enantioselectivity increased to 51 % *ee*. These results indicate that the dual anchoring strategy may be applied to the design of other artificial metalloenzymes to obtain better enantioselective control. It may also be possible in future work to modulate substrate specificity as well.

5.5
Antibodies as Scaffolds for Catalyst Design

5.5.1
Antibodies as Specificity Elements

The immune system of higher eukaryotes produces antibody molecules that recognize a broad range of small molecules. By immunizing animals with specific antigens, produced by conjugating organic compounds to carrier proteins, antibodies that specifically recognize the original small molecule can be generated. This allows protein scaf-

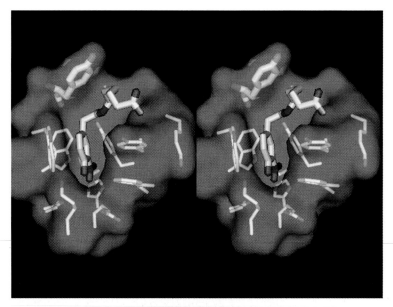

Figure 5.18 Stereo view of the structure of the antibody MOPC 315 complexed with an amide-containing hapten. Only the region near the hapten binding site is shown. The hapten and Y_{34} are shown as larger sticks with the remaining residues in the binding site as smaller sticks. Residues shown include Y_{34}, W_{93} and W_{98} (from V_L) and W_{333}, H_{335}, Y_{339}, R_{350}, K_{359}, Y_{399}, Y_{401} and S_{405} (V_H). Color scheme: Carbon (white), oxygen (red) and nitrogen (blue).

folds that have binding sites tailored for a specific organic molecule to be produced. Figure 5.18 shows the hapten binding site of MOPC 315; an impressive number of specific contacts between the antibody and the nitrophenyl amide containing ligand can be seen. While the immunoglobulin-derived scaffolds are quite large, the ability to create high-affinity substrate binding sites without numerous iterations of design makes this an attractive approach for catalyst design.

5.5.2
Incorporation of an Imidazole Functional Group into an Antibody for Catalysis

The capability of antibodies to provide tailor-made binding sites for catalyst design has been exploited for the preparation of hydrolytic catalysts. The antibody MOPC 315 binds substituted 2,4-dinitrophenyl-containing compounds with association constants ranging from 5×10^4 to 1×10^6 M^{-1} [63]. Imidazole was incorporated into this antibody via a thiol group introduced by chemically modifying K_{52H} in the active site via a disulfide linkage (Figure 5.19). Using a series of coumarin esters as substrates, multiple (>10) hydrolytic turnovers were observed with no loss of activity. The catalytic efficiency, k_{cat}/K_M, for this reaction was over 10^3 times higher than for a model reaction employing 4-methylimidazole.

5.5.3
Comparison of Imidazole-containing Antibodies Produced by Chemical Modification and Site-directed Mutagenesis

Since an imidazole group can be incorporated into a protein through site-directed mutagenesis, such an approach was used to prepare a MOPC 315 mutant containing a histidine residue at the corresponding site. In brief, a hybrid Fv fragment of MOPC 315 was constructed by reconstituting a recombinant variable light chain (V_L) pro-

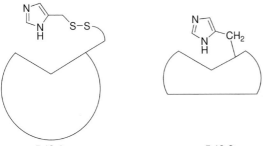

5.19-1	**5.19-2**
Chemically Modified Antibody	Antibody Modified by Mutagenesis

Figure 5.19   Introduction of imidazole into the ligand binding site of MOPC 315 through chemical modification and site-directed mutagenesis.

duced in *E. coli* with a variable heavy chain (V_H) from the antibody [64]. Imidazole was introduced into the combining site by substituting Y_{34} of V_L with His using site-directed mutagenesis (see Y_{34} in Figure 5.18). This His mutant Fv catalyzed the hydrolysis of the 7-hydroxycoumarin esters of 5-(2,4-dinitrophenyl)-aminopentanoic acid 90 000-fold faster than the reaction in the presence of 4-methyl imidazole at pH 7.8. Using that substrate, Fv($Y34H_L$) turned over at least eleven times and retained 44 % of its activity. The loss of activity probably resulted from the accumulation of the inhibitory reaction product. The mutant Fv bound ε-2,4-DNP-L-lysine only six-fold less tightly than wild-type protein, suggesting that the substituted Tyr residue is not involved in the DNP recognition process. Compared to the imidazole-catalytic antibody generated by tethering an imidazole chemically, a sixteen-fold greater rate increase was observed for the reactions catalyzed by His mutant Fv protein under similar conditions. This rate enhancement is likely due to the fewer degrees of the freedom of the imidazole moiety inside the combining site (cf. **5.19-1** and **5.19-2**). While these results alone indicate that the construct prepared by mutagenesis functioned more efficiently than the one obtained from chemical modification, this particular example does not exemplify the most powerful feature of chemical modification, i.e., the ability to incorporate non-natural functionality or spacers into complex protein structures. Considerable use will probably be made of this strategy in future catalyst design.

5.6
Conclusions

The examples described in this chapter reveal the breadth of methods now used to create protein-based catalysts. Structural analysis via X-ray and NMR techniques has proved to be critical for providing atomic structures that serve as the starting point for design. Computer modeling has become a powerful method in both the planning of new designs and in the interpretation of experimental results. Recombinant DNA techniques allow chemists to make mutations or more global changes in protein scaffolds. Together, these tools have allowed researchers to assemble new protein-based catalysts that can be used to probe biological systems or catalyze useful chemical transformations. While these catalysts do not in general rival the efficiency of natural enzymes, impressive rate accelerations have been obtained in some cases. New methods including phage display [65] and antibody production promise to provide greater access to ligand-specific scaffold development. Peptide/protein ligation techniques offer the possibility of increasing the scope of functionality that can be incorporated into protein structures [66]. As the understanding of how enzymes promote chemical reactions with high efficiency increases, so too will the ability to design more efficient protein-based catalysts. Given the enormous power obtained from combining chemical and genetic methods it is likely that the field of protein-based artificial enzyme design will continue to grow in the near future.

References

1 K. E. Neet, D. E. Koshland, Jr., *Proc. Natl. Acad, Sci. U.S.A.* **1966**, *56*, 1606–1611.

2 L. Polgar, M. L. Bender, *J. Am. Chem. Soc.* **1966**, *88*, 3153–3154.

3 Z. P. Wu, D. Hilvert, *J. Am. Chem. Soc.* **1989**, *111*, 4513–4514.

4 H. L. Levine, Y. Nakagawa, E. T. Kaiser, *Biochem. Biophys. Res. Commun.* **1977**, *76*, 64–70.

5 D. R. Corey, P. G. Schultz, *Science* **1987**, *238*, 1401–1403.

6 P. Berglund, M. R. Stabile, M. Gold, J. B. Jones, *Bioorg. Med. Chem. Lett.* **1996**, *6*, 2507–2512.

7 P. Berglund, G. DeSantis, M. R. Stabile, X. Shang, M. Gold, R. R. Bott, T. P. Graycar, T. H. Lau, C. Mitchinson, J. B. Jones, *J. Am. Chem. Soc.* **1997**, *119*, 5265–5266.

8 E. T. Kaiser, D. S. Lawrence, *Science* **1984**, *226*, 505–511.

9 E. T. Kaiser, D. S. Lawrence, S. E. Rokita, *Annu. Rev. Biochem.* **1985**, *54*, 565–595.

10 C. F. Meares, T. G. Wensel, *Acc. Chem. Res.* **1984**, *17*, 202–209.

11 D. S. Sigman, T. W. Bruice, A. Mazumder, C. L. Sutton, *Acc. Chem. Res.* **1993**, *26*, 98–104.

12 I. M. Bell, D. Hilvert, *Perspect. Supramol. Chem.* **1994**, *1*, 73–88.

13 M. D. Distefano, H. Kuang, D. Qi, A. Mazhary, *Curr. Opin. Struct. Biol.* **1998**, *8*, 459–465.

14 J. B. Jones, G. DeSantis, *Acc. Chem. Res.* **1999**, *32*, 99–107.

15 D. Haring, P. Schreier, *Naturwissenschaften* **1999**, *86*, 307–312.

16 G. DeSantis, J. B. Jones, *Curr. Opin. Biotecnol.* **1999**, *10*, 324–330.

17 C.-M. Tann, D. Qi, M. D. Distefano, *Curr. Opin. Chem. Biol.* **2001**, *5*, 696–704.

18 R. Breslow, S. D. Dong, *Chem. Rev.* **1998**, *98*, 1997–2011.

19 D. Qi, C. M. Tann, D. Haring, M. Distefano, *Chem. Rev.* **2001**, *101*, 3081–3111.

20 M. A. Zenkova **2004**, *Artificial Nucleases*, Springer, Germany.

21 C. H. Chen, D. S. Sigman, *Science* **1987**, *237*, 1197–201.

22 M. M. Meijler, I. Zelenko, D. S. Sigman, *J. Am. Chem. Soc.* **1997**, *119*, 1135–1136.

23 T. E. Goyne, D. S. Sigman, *J. Am. Chem. Soc.* **1987**, *109*, 2846–2848.

24 R. Jue, J. M. Lambert, L. R. Pierce, R. R. Traut, *Biochemistry* **1978**, *17*, 5399–406.

25 D. N. Arvidson, C. Bruce, R. P. Gunsalus, *J. Biol. Chem.* **1986**, *261*, 238–243.

26 T. W. Bruice, J. G. Wise, D. S. E. Rosser, D. S. Sigman, *J. Am. Chem. Soc.* **1991**, *113*, 5446–5447.

27 D. H. Ohlendorf, W. F. Anderson, R. G. Fisher, Y. Takeda, B. W. Matthews, *Nature* **1982**, *298*, 718–723.

28 R. G. Brennan, Y. Takeda, J. Kim, W. F. Anderson, B. W. Matthews, *J. Mol. Biol.* **1986**, *188*, 115–118.

29 R. G. Brennan, S. L. Roderick, Y. Takeda, B. W. Matthews, *Proc. Natl. Acad. Sci. U.S.A.* **1990**, *87*, 8165–8169.

30 P. Leighton, P. Lu, *Biochemistry* **1987**, *26*, 7262–7271.

31 G. Xiao, D. L. Cole, R. P. Gunsalus, D. S. Sigman, *Protein Sci.* **2002**, *11*, 2427–2436.

32 C. Q. Pan, J. A. Feng, S. E. Finkel, R. Landgraf, D. Sigman, R. C. Johnson, *Proc. Natl. Acad. Sci. U.S.A.* **1994**, *91*, 1721–1725.

33 D. Kostrewa, J. Granzin, C. Koch, H. W. Choe, S. Raghunathan, W. Wolf, J. Labahn, R. Kahmann, W. Saenger, *Nature* **1991**, *349*, 178–180.

34 H. S. Yuan, S. E. Finkel, J. A. Feng, M. Kaczor-Grzeskowiak, R. C. Johnson, R. E. Dickerson, *Proc. Natl. Acad. Sci. U.S.A.* **1991**, *88*, 9558–9562.

35 R. H. Ebright, Y. W. Ebright, P. S. Pendergrast, A. Gunasekera, *Proc. Natl. Acad. Sci. U.S.A.* **1990**, *87*, 2882–2886.

36 R. H. Ebright, Y. W. Ebright, A. Gunasekera, *Nucl. Acids Res.* **1989**, *17*, 10 295–10 305.

37 P. S. Pendergrast, Y. W. Ebright, R. H. Ebright, *Science* **1994**, *265*, 959–962.

38 S. C. Schultz, G. C. Shields, T. A. Steitz, *Science* **1991**, *253*, 1001–1007.

39 J. A. Bown, J. T. Owens, C. F. Meares, N. Fujita, A. Ishihama, S. J. W. Busby, S. D. Minchin, *J. Biol. Chem.* **1999**, *274*, 2263–2270.

40 T. D. Tullins, B. A. Dombroski, *Proc. Natl. Acad. Sci. U.S.A.* **1986**, *83*, 5469–5473.

41 C. Liu, M. Wang, T. Zhang, H. Sun, *Coord. Chem. Rev.* **2004**, *248*, 147–168.

42 L. Banaszak, N. Winter, Z. Xu, D. A. Bernlohr, S. Cowan, T. A. Joner, *Adv. Protein Chem.* **1994**, *45*, 89–151.

43 M. E. Hodsdon, J. W. Ponder, D. P. Cistola, *J. Mol. Biol.* **1996**, *264*, 585–602.

44 H. Kuang, M. L. Brown, R. R. Davies, E. C. Young, M. D. Distefano, *J. Am. Chem. Soc.* **1996**, *118*, 10 702–10 706.

45 J. J. Ory, A. Mazhary, H. Kuang, R. R. Davies, M. D. Distefano, L. J. Banaszak, *Protein Eng.* **1998**, *11*, 253–261.

46 H. Kuang, R. R. Davies, M. D. Distefano, *Bioorg. Med. Chem. Lett.* **1997**, *7*, 2055–2060.

47 H. Kuang, M. D. Distefano, *Protein Eng.* **1997**, *10*, 25.

48 H. Kuang, M. D. Distefano, *J. Am. Chem. Soc.* **1998**, *120*, 1072–1073.

49 D. Qi, H. Kuang, M. D. Distefano, *Bioorg. Med. Chem. Lett.* **1998**, *8*, 875–880.

50 H. Kuang, D. Haring, D. Qi, A. Mazhary, M. D. Distefano, *Bioorg. Med. Chem. Lett.* **2000**, *10*, 2091–2095.

51 D. Haring, H. Kuang, L. J. Banaszak, M. D. Distefano, *Protein Eng.* **2002**, *15*, 603–610.

52 D. Haring, M. D. Distefano, *Bioconj. Chem.* **2001**, *12*, 385–390.

53 D. Haring, M. D. Distefano, *Bioorg. Med. Chem.* **2001**, *9*, 2461–2466.

54 K. Kim, D. P. Cistola, C. Frieden, *Biochemistry* **1996**, *35*, 7553–7558.

55 R. A. Steele, D. A. Emmert, J. Kao, M. E. Hodsdon, C. Frieden, D. P. Cistola, *Protein Sci.* **1998**, *7*, 1332–1339.

56 D. Qi **2003**, *Chemistry PhD Thesis,* University of Minnesota, 140.

57 R. R. Davies, M. D. Distefano, *J. Am. Chem. Soc.* **1997**, *119*, 11 643–11 652.

58 L. M. Sayre, K. V. Reddy, A. R. Jacobson, W. Tang, *Inorg. Chem.* **1992**, *31*, 935–937.

59 R. R. Davies, H. Kuang, D. Qi, A. Mazhary, E. Mayaan, M. D. Distefano, *Bioorg. Med. Chem. Lett.* **1999**, *9*, 79–84.

60 C-M. Tann **2003**, *Chemistry PhD Thesis,* University of Minnesota, 201.

61 M. Ohashi, T. Koshiyama, T. Ueno, M. Yanase, H. Fujii, Y. Watanabe, *Angew. Chem. Int. Ed.* **2003**, *42*, 1005–1008.

62 J. R. Carey, S. K. Ma, T. D. Pfister, D. K. Garner, H. K. Kim, J. A. Abramite, Z. Wang, Z. Guo, Y. Lu, *J. Am. Chem. Soc.* **2004**, *126*, 10 812–10 813.

63 S. J. Pollack, P. G. Schultz, *J. Am. Chem. Soc.* **1989**, *111*, 1929–1931.

64 E. Baldwin, P. G. Schultz, *Science* **1989**, *245*, 1104–1107.

65 R. H. Hoess, *Chem. Rev.* **2001**, *101*, 3205–3218.

66 P. E. Dawson, T. W. Muir, I. Clark-Lewis, S. B. H. Kent, *Science* **1994**, *266*, 776–779.

6
Artificial Hydrolytic Metalloenzymes

Jik Chin and Hae-Jo Kim

6.1
Introduction

Hydrolytic metalloenzymes [1–3] play many important roles in life. Proteases, phospholipases and nucleases that hydrolyze proteins, lipids and nucleic acids often require one or more metal ions. Metalloenzymes that hydrolyze phosphoproteins play important roles in signal transduction [4–7]. Enzymes that make DNA by transesterification such as polymerases [8] and reverse transcriptases [9] are activated by two metal ions. In addition, many RNA enzymes that catalyze transesterification and hydrolysis reactions are activated by more than one metal ion [10]. Over the years, there has been considerable interest in understanding the role of metal ions in these important biological reactions. Much has been learned through elegant designs and careful analyses of simple metalloenzyme models [11–15]. These studies provide valuable information on how some of these enzymes provide enormous rate-accelerations for the hydrolysis reactions. Furthermore, the knowledge gained from such studies may provide interesting insights into designing metal complexes that catalyze not only hydrolysis reactions but also a wide variety of organic reactions. In This chapter presents some of the fundamental roles that metal ions play in hydrolyzing esters, amides, nitriles and phosphate esters.

6.2
Reactivity of Substrates

To understand the role of metal ions in hydrolysis reactions, it is useful to first consider the 'background' hydrolysis reactions. Table 6.1 lists the second-order rate constants for hydroxide-catalyzed hydrolysis of various substrates. The reactivity of methyl acetate (first entry in Table 6.1) [16] is comparable to those of other unactivated esters found in nature (e.g. acetyl choline and carboxyl esters in phospholipids). The reactivity of N-methylacetamide (second entry in Table 6.1) [17] is comparable to those of typical peptides (1.1×10^{-6} M^{-1} s^{-1}) [18] and that of dimethyl phosphate (P–O bond

Artificial Enzymes. Edited by R. Breslow
Copyright © 2005 WILEY-VCH Verlag GmbH & Co. KGaA, Weinheim
ISBN: 3-527-31165-3

Table 6.1 Second-order rate constant (k_{OH}) for hydroxide-catalyzed hydrolysis of substrates at 25 °C.

Substrate	k_{OH} (M^{-1} s^{-1})
$CH_3C(O)OCH_3$	1.5×10^{-1}
$CH_3C(O)NHCH_3$	5.6×10^{-6}
CH_3CN	1.6×10^{-6}
$P(O)(OCH_3)_3$	1.6×10^{-4}
$NaOP(O)(OCH_3)_2$	6.8×10^{-12}

cleavage) [19] is about the same as typical phosphodiester bonds in DNA and phospholipids [13]. Not surprisingly, nature chose to preserve the genetic material using the linkage that is by far the most difficult to hydrolyze.

Notably, although the hydroxide rate for methyl acetate hydrolysis is many orders of magnitude greater than that for N-methyl acetamide, the water rate for ester hydrolysis (3.16×10^{-10} s^{-1} at 25 °C) [20] is only a few times greater than the water rate for typical peptide hydrolysis (10^{-10} s^{-1}) [18]. It may seem somewhat surprising that the rates of hydrolysis of esters at neutral pH are not much greater than those of amides. However, the leaving group amine in amide hydrolysis is protonated at neutral pH while the corresponding ether oxygen in ester hydrolysis is not. Thus, amides and esters have comparable reactivities at neutral pH, but esters are far more reactive than amides under basic conditions.

As discussed in this chapter, there are three direct modes of activation that metal ions can provide for hydrolysis reactions, i.e., by Lewis acid, nucleophile and leaving group. In addition, metal coordinated water molecules can act as general acid catalysts and metal coordinated hydroxides can act as general base catalysts. We are mainly concerned with the three direct modes of activation since there is no particular advantage to using metal-based general acids and bases.

6.3
Lewis Acid Activation

Metal ions can increase the electrophilicity of the substrates by direct coordination (Figure 6.1). Lewis acid activation is important not only for hydrolysis of esters, amides, nitriles and phosphates [11–15] but also for other organic reactions, including epoxide opening, aldol condensation, Michael addition, reduction and Diels–Alder reactions (Figure 6.1). Indeed, chiral Lewis acids have been used as catalysts to carry out a wide range of stereoselective transformations [21–25]. In all these reactions, the role of the metal ion is to stabilize the developing negative charge on the coordinated atom. Often, the coordinated atom is oxygen and thus the role of the metal ion is mostly oxyanion stabilization.

One way to estimate the relative Lewis acidity of metal ions is by considering the Brønsted acidity of the metal coordinated water molecules. Table 6.2 lists the pK_a of

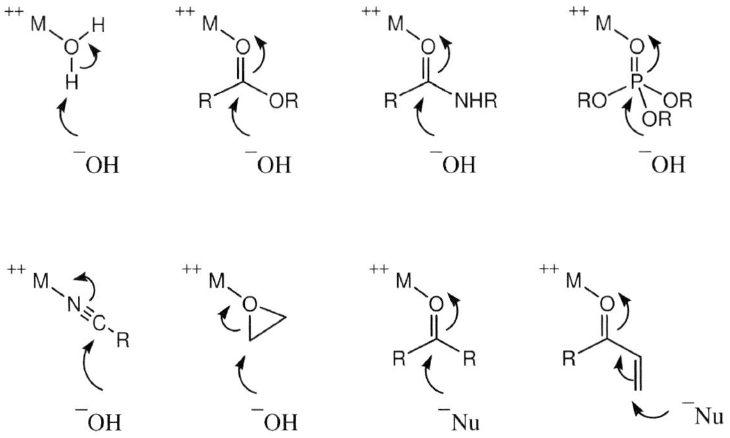

Figure 6.1 Lewis acid activation for hydrolysis and other organic reactions.

some metal-bound water molecules [26]. Metal cations increase the acidity of the bound water molecule by stabilizing the metal-bound hydroxide anion. Here again, the role of the metal ion is oxyanion stabilization (Figure 6.1). Calcium ions increase the acidity of the free water molecule almost a thousand-fold while zinc ions increase the acidity almost ten million-fold. The stronger the Lewis acidity of the metal ion, the lower is the pK_a of the bound water molecule (Table 6.2). Hydronium, or water bound to proton, is far more acidic than water bound to Zn(II) [water bound to highly Lewis acidic metal ions such as Ti(IV) is expected to be more acidic than hydronium]. Protonated substrates are expected to be far more reactive to hydrolysis than the corresponding substrates coordinated to metal ions in Table 6.2. However, metal ions can provide greater activation than proton through cooperative interaction of Lewis acid activation and nucleophile activation.

The active site of bovine carbonic anhydrase consists of a tetrahedrally coordinated zinc ion (1) [27] with a coordinated water molecule whose pK_a (7.5) [28, 29] is con-

Table 6.2 Acidity of free and metal ion or proton bound water molecules.

Water species	pK_a
H_2O	15.72
Ca(II)–OH$_2$	12.85
Mg(II)–OH$_2$	11.44
Mn(II)–OH$_2$	10.59
Co(II)–OH$_2$	10.2
Ni(II)–OH$_2$	9.86
Fe(II)–OH$_2$	9.5
Zn(II)-OH$_2$	8.96
H_3O^+	−1.72

siderably lower than that of water bound to free zinc ion (8.96). There has been much interest in developing simple zinc hydrate complexes with neutral pK_as. Woolley reported a five-coordinate zinc hydrate complex (**2**) with a pK_a of 8.7 [30, 31]. This complex was shown to be active for catalyzing the hydration of carbon dioxide. Kimura et al. reported a tetrahedral zinc hydrate complex (**3**) with a pK_a (7.3) [32] that closely matches that of the zinc hydrate at the active site of carbonic anhydrase. Kimura's complex was a good catalyst for hydrolyzing esters with good or poor leaving groups.

Ligands can be designed to influence the acidity of the metal center in several ways. Metal ions with fewer coordinated ligands should be more acidic (Figure 6.2). Thus, tetrahedral zinc should be more acidic than five- or six-coordinate zinc. Decreasing the basicity of the ligand should increase the Lewis acidity of the metal center. Electron-withdrawing groups attached to ligands should, therefore, enhance the acidity of the metal. Steric effects or strain could also be used to weaken the metal–ligand interaction. The equilibrium constant for binding of the twelve-membered ring macrocycle to zinc (**3**) is about a hundred times smaller than that for binding of the eleven-membered ring macrocycle to zinc (**4**) [32]. The hundred-fold weaker binding of the triaza ligand in **3** appears to translate to a ten-fold increase in acidity of the metal-bound water (pK_a of **4** is 8.2). While tight binding of the ligand to the metal is desirable to form a stable complex, weak binding is needed to enhance the Lewis acidity of the metal. In **3** and **4**, the equilibrium constant for binding of the macrocyclic ligand to the zinc ion is large (when compared to the binding of linear ligands to the metal ion) yet the three nitrogens in **3** are held sufficiently apart from the metal by strain. The weak interaction between the nitrogens and zinc ion in **3** results in increased acidity of the metal ion.

Hydrolytic metalloenzymes can provide enormous rate accelerations for hydrolyzing a wide variety of substrates [1–3]. It is an interesting challenge to be able to determine how much rate acceleration is possible due to Lewis acid activation alone. An upper limit to the rate acceleration due to Lewis acid activation could be obtained from the change in acidity of water upon coordination to a metal ion (Table 6.1). For example, if a metal ion increases the acidity of the water molecule by a million-fold, this metal ion should provide up to a maximum of a million-fold rate acceleration for hydrolysing esters (or amides, nitriles or phosphates; Figure 6.1). A greater charge buildup on the coordinated oxygen is anticipated on going from the metal-bound water

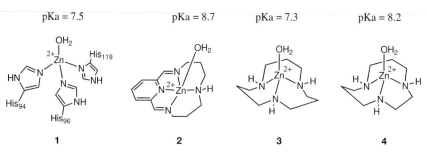

Figure 6.2 Acidity of zinc-bound water.

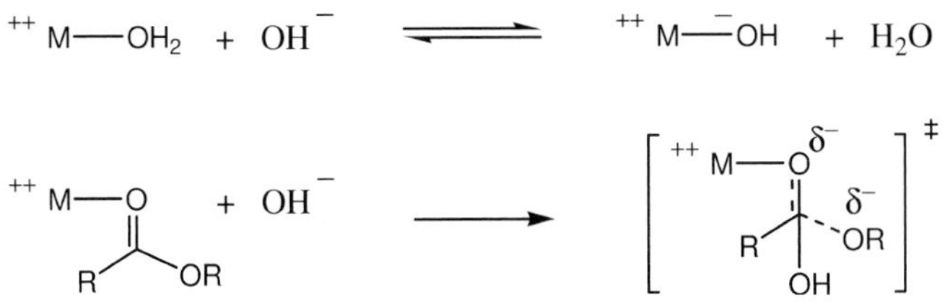

$$^{++}M\text{—}OH_2 + OH^- \rightleftharpoons {}^{++}M\text{—}OH^- + H_2O$$

Figure 6.3   Charge build-up on coordinated oxygen.

to the metal-bound hydroxide when compared to the charge buildup on going from the metal-bound ester to the metal-bound transition state for the hydrolysis reaction (Figure 6.3). Assuming a linear relationship between energy and charge, it should be possible to estimate the effect of metal ions on the rate of hydrolysis reactions. If the net change in charge (on the metal coordinated oxygen) due to the hydrolysis reaction at the transition state is 50 % of that due to the deprotonation reaction, a metal ion that provides a million-fold increase in the acidity of a water molecule should give a thousand-fold increase in the rate of hydrolysis. Lewis acid catalysis can be compared to buffer catalysis. In general acid catalysis, increasing the acidity of the catalyst ten-fold increases the rate of the catalyzed reaction by a maximum of ten-fold (Brønsted coefficient $\alpha = 1$). If the extent of proton transfer at the transition state of the general acid catalyzed reaction is 50 %, increasing the acidity of the catalyst a million-fold should result in a thousand-fold increase in the rate of the catalyzed reaction [33, 34].

Substitutionally inert Co(III) or Ir(III) complexes have been used to measure directly the effect of Lewis acid activation on the hydrolysis of an amide [35–37], a nitrile [38] and a phosphate triester [39] (Figure 6.4). The pK_a of the cobalt-bound water molecule in **5** is 6.6 [40]. Thus the upper limit for the rate-acceleration due to Lewis acid activation with this metal in the hydrolysis of esters, amides, nitriles and phosphates should be close to 10^9-fold. Although the observed rate accelerations for the hydrolysis reac-

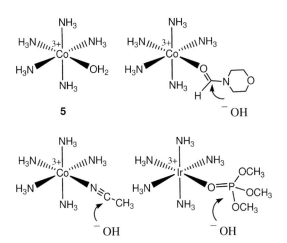

Figure 6.4   Hydrolysis of Co(III) complex bound substrates with hydroxide.

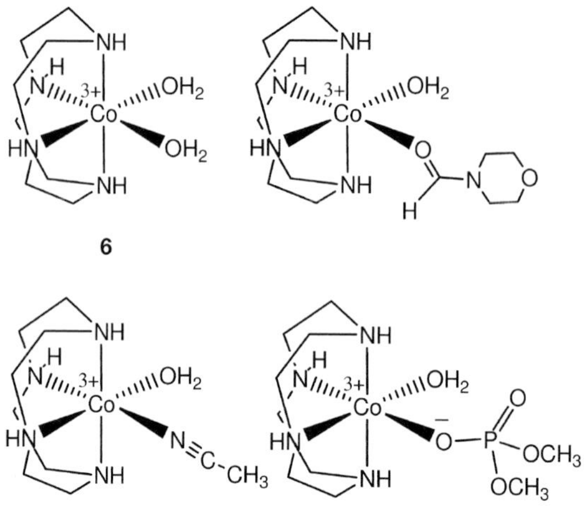

Figure 6.5 Equilibrium coordination of substrates to **6**.

tions are quite remarkable, they fall considerably short of the upper limit. With formyl morpholine, acetonitrile and trimethyl phosphate, the rate accelerations due to Lewis acid activation are 2×10^5-, 2×10^6- and 4×10^2-fold respectively. Other metal ions that are less Lewis acidic (Table 6.2) than Co(III) should provide somewhat smaller rate accelerations for the hydrolysis reactions.

The efficiency of Lewis acid activation depends not only on the reactivity of the bound substrate but also on the equilibrium constant for coordination of the substrate. The equilibrium constants for binding of an amide, a nitrile and phosphates to Co(III) complexes have been measured (Figure 6.5). Formyl morpholine binds to **6** with an equilibrium constant of 0.4 M^{-1} [35]. Binding of acetamide to **6** could not be detected. The steric effect of the methyl group is expected to significantly lower the binding of acetamide compared with that of formyl morpholine.

The equilibrium constant for binding of acetonitrile to **6** (2.5 M^{-1}) [41] is comparable to that for binding of formyl morpholine to **6** even though acetonitrile is about ten orders of magnitude less basic than the amide [42]. The tight binding of acetonitrile may be in part due to the lack of steric interactions between the nitrile and the metal complex. In nitrile coordination, the sp-orbital on the nitrogen is expected to interact with the d-orbital on the metal in a linear arrangement. In amide coordination, the sp^2-orbital on the carbonyl oxygen is expected to interact with the d-orbital on the metal. In the coordinated complex, the formyl hydrogen should be positioned towards the metal complex since it is smaller than the amide nitrogen.

The equilibrium constant for binding dimethyl phosphate (2.8 M^{-1}) [43] to **6** is greater than those for binding formyl morpholine and acetonitrile to **6**. Favorable electrostatic interaction between the metal ion and the coordinating oxygen of dimethyl phosphate appears to counter any possible steric interactions. The exact value of the equilibrium constants for binding of substrates to metal ions and complexes will depend on the substrate, the metal and the ligand surrounding the metal. The examples in Figure 6.5

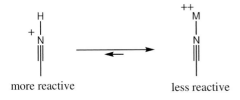

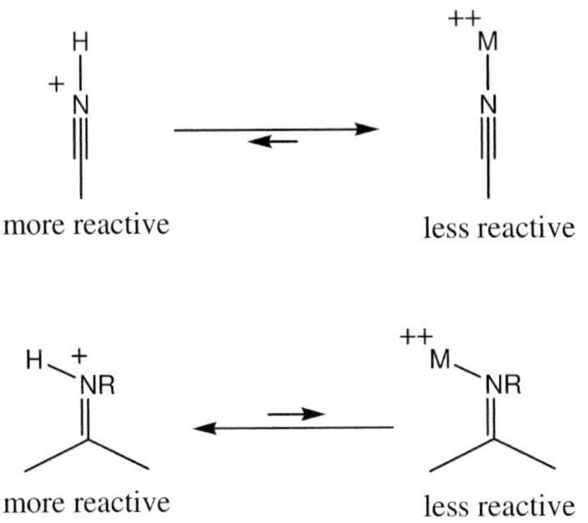

Figure 6.6 Protonation vs. metal coordination.

show that a range of substrates can bind reasonably well to metal complexes even when solvent water molecules can compete for coordination to the metal.

Lewis acid activation works particularly well for weakly basic substrates that bind well to metal ions (Figure 6.6). Nitriles are excellent substrates for Lewis acid activation since they are weakly basic (pK_a of protonated acetonitrile is about −10) [42] yet they bind well to metal ions (binding constant for acetonitrile to **6** is 2.5 M^{-1}). Imines are about fifteen orders of magnitude more basic than nitriles. Thus, in neutral aqueous solutions, significant amounts of imines are protonated. Since protonated substrates are generally more reactive than metal coordinated ones (Table 6.1), Lewis acid activation of imines with metal ions in aqueous solutions is difficult.

In addition to the above hydrolysis reactions, Lewis acid activation has been used to carry out a range of stereoselective transformations. Many interesting chiral Lewis acids have been developed over the past decade [21–25]. A single chiral Lewis acid (**7**) has been shown to catalyze a wide variety of carbon–carbon bond forming reactions (Figure 6.7) [22]. There is much current interest in developing such privileged catalysts [21]. Solvent effects play an important role in Lewis acid activation. Both binding of substrates and stabilization of oxyanionic transition states may be enhanced in organic solvents.

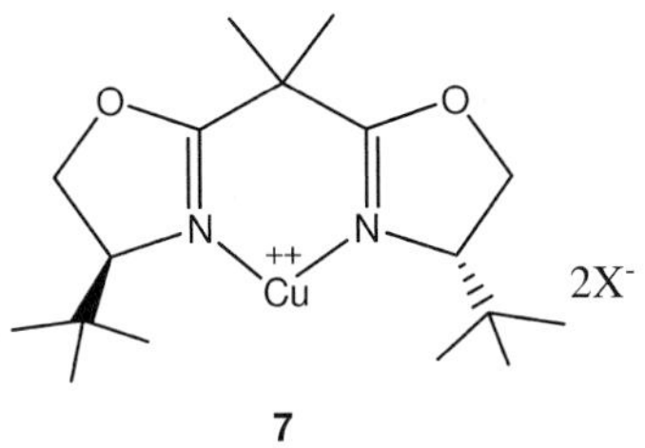

7

Figure 6.7 A privileged chiral Lewis acid.

6.4
Nucleophile Activation

Metal hydroxides are thought to be involved in various hydrolytic metalloenzymes, including carbonic anhydrase, carboxypeptidase, and nucleases [1–3]. Simple metal hydroxides are active as nucleophilic catalysts for the hydration of carbon dioxide and the hydrolysis of acetic anhydride and p-nitrophenyl acetate [31]. The rate of these reactions is accelerated with increase in basicity of the metal hydroxide. The Brønsted equation has been used to show the linear relationship between the pK_bs of the metal hydroxide and the logarithms of the rate constants for the hydration of carbon dioxide ($\beta_{nuc} = 0.15$) and the hydrolysis of acetic anhydride ($\beta_{nuc} = 0.25$) and p-nitrophenyl acetate ($\beta_{nuc}= 0.35$). The Brønsted coefficients (β_{nuc}) can be used to approximate the extent of bond formation between the metal hydroxide and the substrate at the transition state (Figure 6.8). With joint Lewis acid activation and metal hydroxide activation, the extent of charge buildup on the carbonyl oxygen (Figure 6.3) should be related to the extent of bond formation between the metal hydroxide and the substrate (Figure 6.8).

A metal hydroxide that is well positioned next to the substrate could in principle provide enormous rate acceleration for the hydrolysis reaction. If the metal hydroxide is positioned such that the entropic barrier for hydrolysis is eliminated without affecting the enthalpic barrier [44], the effective concentration of the metal hydroxide is expected to be about 10^8 M [45]. The local concentration of this nucleophile would be 10^{15} ($10^8/10^{-7}$) times greater than the background hydroxide concentration at neutral pH. The basicity and nucleophilicity of a metal hydroxide is lower than those of the metal free hydroxide. For a hydrolysis reaction with a β_{nuc} of 0.5, a metal hydroxide with a pK_b of 7.0 would be about $10^{4.4}$ times less reactive than hydroxide. Thus, a metal hydroxide that is $10^{4.4}$ times less reactive than hydroxide could still provide over a 10^{10}-fold rate acceleration (over the background hydroxide rate) for the hydrolysis reaction if the effective molarity of the metal hydroxide is 10^8 M. In **8** [46], **9** [47] **10** [48] and **11** [49] the well-positioned metal hydroxides provide enormous rate accelerations (10^7 to 10^9) for hydration of the nitriles and hydrolysis of the amides (Figure 6.9). In general, effective molarities of well-positioned intramolecular general base groups are much lower (~1 M) than those of intramolecular nucleophiles [45]. Consequently, there is little advantage to using intramolecular metal hydroxides as general base catalysts when compared with utilizing external general bases.

Figure 6.8 Metal hydroxide attack.

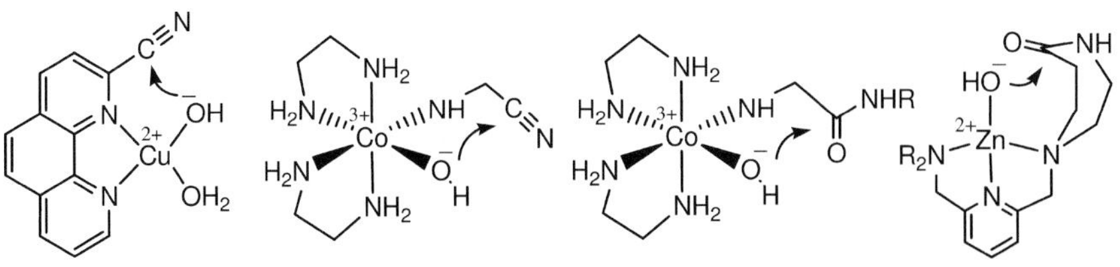

Figure 6.9 Metal hydroxides in nitrile hydration (**8** and **9**) and amide hydrolysis (**9** and **10**).

6.5
Leaving-group Activation

Leaving-group activation could be considered to be the reverse of nucleophile activation. In ester hydrolysis, metal ions can activate the leaving group by coordinating to the leaving group oxygen and lowering the basicity of the alkoxide [50, 51]. When strong nucleophiles (e.g. hydroxide or alkoxide) are used to cleave esters with good leaving groups (e.g. *p*-nitrophenol or acetate), the reaction takes place rapidly and the Brønsted coefficient (β_{lg}) is about 0.3 [52]. In such cases, lowering the basicity of the leaving group by 10 log units through metal coordination will result in about a thousand-fold increase in the cleavage reaction rate. When weak nucleophiles (e.g. acetate or phenoxide) are used to cleave esters with poor leaving groups (e.g. alkoxide), the reaction is slow and β_{lg} approaches 1.0 [52]. In this case, a ten-fold decrease in basicity of the leaving group will result in a matching increase in the rate of hydrolysis.

In amide hydrolysis, the leaving group amines are generally basic enough to be protonated at neutral pH (Figure 6.10). Leaving-group activation by metal coordination is not important for amide hydrolysis since protonated amines are stronger acids (and better leaving groups) than metal coordinated amines. In general, groups that are easily protonated will not be activated by metal ions (Figures 6.6 and 6.10).

For the transesterification and hydrolysis of phosphate diesters, β_{lg} ranges from about 0.67 to 0.76 [53, 54]. Thus, leaving-group activation by metal coordination should result in a large rate acceleration ($\sim 10^7$–10^8-fold) for both transesterification of RNA and hydrolysis of DNA. However, it is a challenge to design metal complexes that will bind well enough to the leaving group oxygen of carboxyl and phosphate esters.

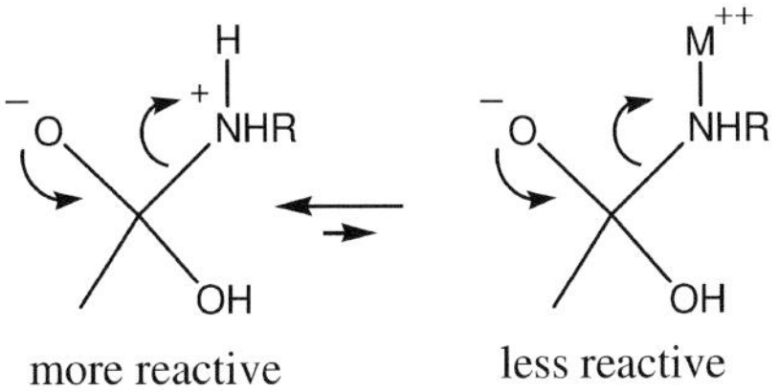

more reactive   less reactive

Figure 6.10 Tetrahedral intermediate for amide hydrolysis.

6.6
Combining Lewis Acid Activation and Nucleophile Activation

Lewis acid activation (Figure 6.3) and nucleophile activation (Figure 6.8) may be combined with one or more metal complexes to provide large rate accelerations for hydrolysis and other organic reactions. While Lewis acid activation alone (Figure 6.4) provides about 2×10^5-fold rate acceleration for the hydrolysis of formyl morpholine, a combination of Lewis acid activation and nucleophile activation (12) gives about 4×10^7-fold rate acceleration for the hydrolysis reaction over the background hydroxide rate at pH 7 [35]. For acetonitrile, combining Lewis acid activation and nucleophile activation (13) gives about a 10^{10}-fold rate acceleration for the hydration reaction [41]. Thus, the combined activation provides an almost 10^4-fold greater rate acceleration than Lewis acid activation alone (Figure 6.4). For phosphate diesters, Lewis acid activation gives a ca. 50-fold rate acceleration for the hydrolysis reaction [55]. The combination of Lewis acid activation and nucleophile activation (14) gives about a 10^{10}-fold rate acceleration for the hydrolysis reaction (over the background hydroxide rate at pH 7) [43, 54]. Assuming Lewis acid and nucleophile activations are additive for each reaction, nucleophile activation alone would provide 2×10^2-, 5×10^3-, and 2×10^8-fold rate accelerations for the hydrolysis of formyl morpholine, acetonitrile and dimethyl phosphate, respectively. The greater effect of nucleophile activation for phosphate diester hydrolysis may be due in part to the longer P–O (14) bonds compared with C–O (12) or C–N (13). The longer bond should result in less strain in the four-membered ring transition state (Figure 6.11).

Interestingly, catalytic turnover can be achieved for the hydration of nitriles with the Co(III) complex (13). Hence, fundamental information gained from studying intramolecular model systems (8–11) that do not give catalytic turnover can be applied to obtain true catalysts. Lewis acid activation and nucleophile activation can also be combined using mononuclear Co(III) [56], Cu(II) [51] or Zn(II) [32] complexes to hydrolyze unactivated esters with catalytic turnover.

The combination of Lewis acid and nucleophile activations in 12–14 requires four-membered ring transition states. Interestingly, the mechanism for carboxypeptidase A (CPA) catalyzed hydrolysis of peptides also appears to involve joint Lewis acid and nucleophile activations that lead to the formation of a four-membered ring transition state. Christianson and Lipscomb [57] have determined the crystal structure (15) of a ketone bound to CPA. Surprisingly, the ketone is in its hydrated form with both oxygens of the gem-diol bound to the active-site zinc of CPA (Figure 6.12).

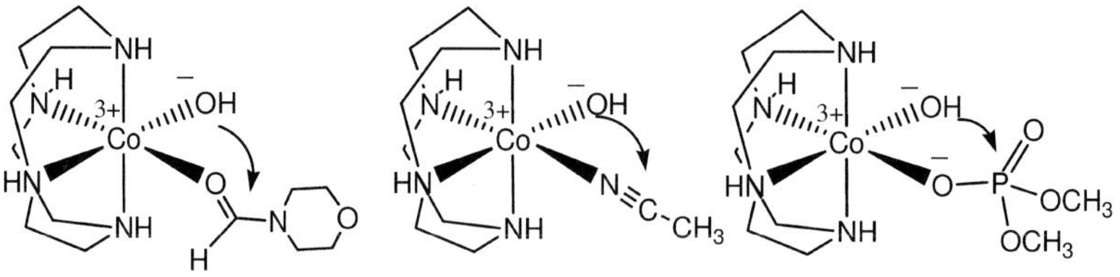

12 **13** **14**

Figure 6.11 Joint Lewis acid activation and nucleophile activation.

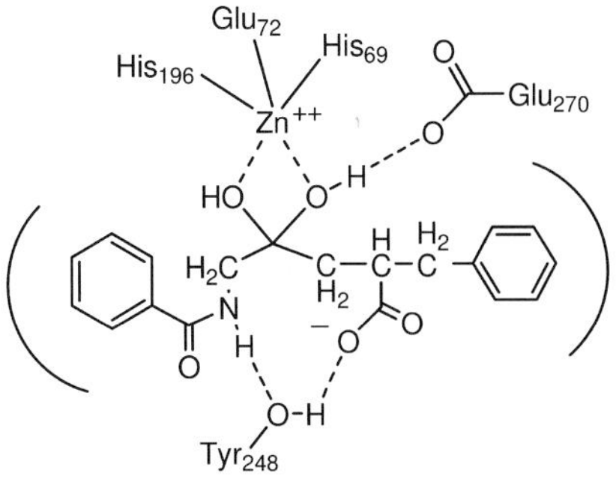

Figure 6.12 Four-membered ring transition state analog at the active site of carboxypeptidase A.

Dinuclear metal complexes have been used to combine Lewis acid and nucleophile activations. For nitrile hydration, Lewis acid activation alone can provide a ca. 10^6-fold rate acceleration (Figure 6.4) and a well-positioned metal hydroxide can provide another 10^9-fold rate enhancement (**8**, **9**). Sargeson et al. [58] found that joint Lewis activation and nucleophile activation with a dinuclear Co(III) complex (**16**) provides an amazing 10^{15}-fold rate acceleration over the background hydroxide rate for the hydration reaction (Figure 6.13). Thus, Lewis acid and hydroxide activations appear to be additive ($10^{6+9} = 10^{15}$). The greater reactivity of the dinuclear system (**16**) over the mononuclear system (**13**) is probably due, mainly, to the better positioning of the metal hydroxide in **16**. In **13**, hydration is accompanied by formation of a strained four-membered ring transition state, whereas in **16** nucleophilic attack of the metal hydroxide on the coordinated nitrile affords a six-membered ring transition state. Rapid hydrolysis of an amide by joint Lewis acid and nucleophile activations has also been accomplished with a dinuclear metal complex (Figure 6.13, **17**) [59].

There has been prolonged interest in developing catalysts or reagents that cleave protein molecules sequence specifically for their potential in the study of proteins as well as in the generation of a new class of therapeutic agents [60–62]. Sequence specific hydrolysis of proteins and oligopeptides has been achieved by combining Lewis acid activation and nucleophile activation using Pd(II) [63] and Co(III) complexes [64].

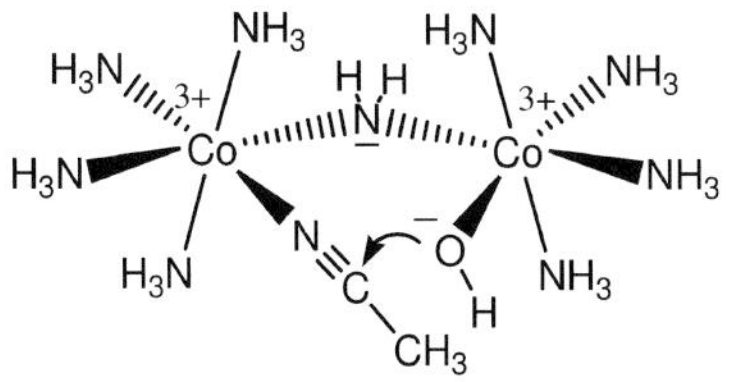

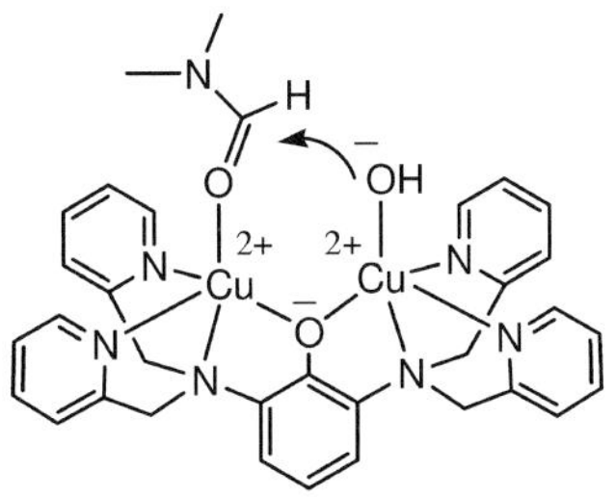

Figure 6.13 Joint Lewis acid activation and nucleophile activation with dinuclear metal complexes.

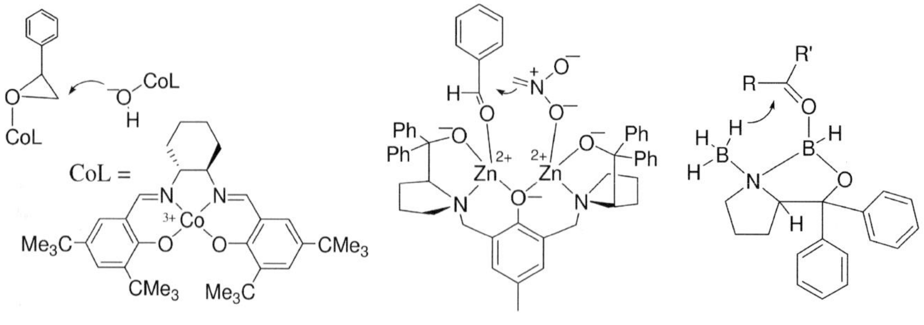

Figure 6.14 Lewis acid and nucleophile activation in stereoselective reactions.

In addition to the above hydrolysis reactions, dinuclear approach to providing joint Lewis acid activation and nucleophile activation has been applied to other organic reactions (Figure 6.14) including stereoselective ring-opening of epoxides (**18**) [65, 66], stereoselective aldol condensation (**19**) [67, 68], and stereoselective reduction (**20**) reactions [69].

6.7
Double Lewis Acid Activation

Single Lewis acid activation provides substantially greater rate acceleration for the hydrolysis of carboxyl esters [70], amides [35–37], and nitriles [38] than for the hydrolysis of phosphate diesters [55]. The rates for hydroxide-catalyzed hydrolysis of esters, amides and nitriles coordinated to Co(III) complexes are 10^4- to 10^6-fold greater than those for hydroxide-catalyzed hydrolysis of the corresponding free substrates (Figure 6.4). Sargeson's research team showed that the rate of hydroxide-catalyzed hydrolysis of a phosphate diester that is coordinated to a Co(III) complex is only about 50 times greater than that of the uncoordinated diester [55]. They also showed that the hydroxide rate for the hydrolysis of a phosphate triester that is coordinated to a Co(III) complex is about 400 times greater than that for the hydrolysis of the uncoordinated triester [39]. Unfortunately, single Lewis acid activation provides the least rate acceleration for the most stable substrate (Table 6.1). The developing negative charge in the single Lewis acid activated phosphate diester cleavage reaction (Figure 6.15) is not stabilized by direct coordination (**21**) as it is in the single Lewis acid activated phosphate triester hydrolysis (**22**) (or in the single Lewis acid activated ester, amide or nitrile hydrolysis). The developing negative charge in phosphate diester hydrolysis could be stabilized by coordinating both phosphoryl oxygens to one or two metal ions (**23**).

Phosphates bridge dinuclear metal centers in the active sites of some phosphoesterases [1–3]. Dinuclear metal complexes should be able to provide double Lewis acid activation for hydrolyzing phosphates. To quantify double Lewis acid activation for cleaving phosphate diesters, we studied the reaction of **24** (Figure 6.16), which has

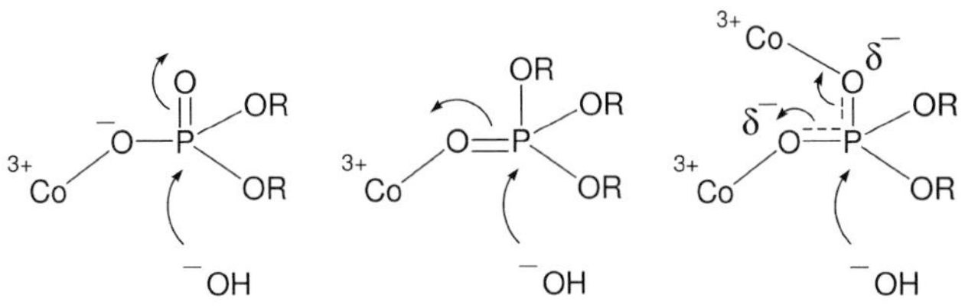

Figure 6.15 Single and double Lewis acid activation in phosphate hydrolysis.

two substitutionally inert Co(III) centers [71–73]. The second-order rate constant for hydroxide-catalyzed transesterification of **24** is 4.3×10^2 M^{-1} s^{-1} at 25 °C. By comparison, the second-order rate constant for hydroxide-catalyzed transesterification of the free diester in solution is 9.8×10^{-4} M^{-1} s^{-1} at 25 °C [74]. Remarkably, double Lewis acid activation in the model system provides a ca. 4×10^5-fold rate enhancement for cleaving the diester, fairly independent of the leaving group basicity. The rate acceleration due to double Lewis acid activation in **24** is significantly greater than the square of the rate acceleration due to single Lewis acid activation for phosphate diester cleavage (50^2). Thus, there appears to be cooperativity between the two metal centers in double Lewis acid activation for phosphate diester transesterification. The rate acceleration may in part be due to strain in the O–P–O bond – the X-ray structure reveals that the bond angle (117.4°) is considerably larger than the tetrahedral value (109.5°) and close to what it should be in the trigonal bipyramidal transition state (120°).

In addition to the remarkable effects on reactivity, there is significant cooperativity between the two metal centers in the dinuclear Co(III) complex (**24**) for coordinating phosphate diesters. The equilibrium constant for monodentate coordination of dimethyl phosphate to a mononuclear Co(III) complex (Figure 6.5) is only about 2.8 M^{-1} [43], but is >330 M^{-1} for bridging dimethyl phosphate to the dinuclear Co(III) center in **24** [75].

Although Co(III) complexes are very useful for quantifying the effects of different types of activation, it is generally difficult to obtain catalytic turnover because they are

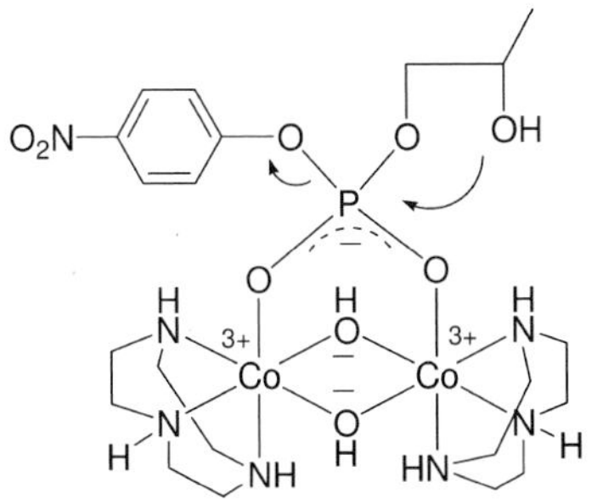

Figure 6.16 Quantifying double Lewis acid activation.

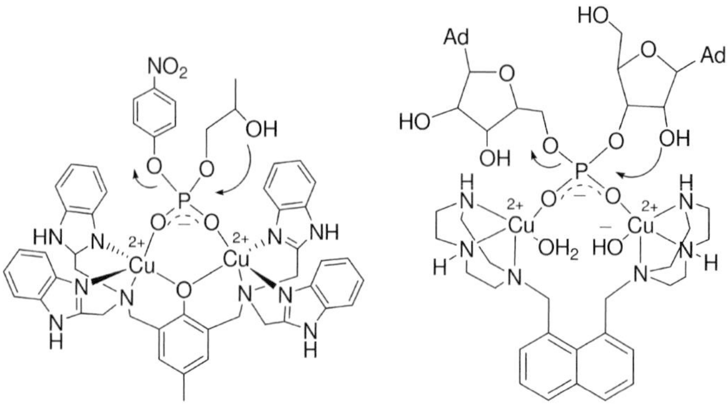

Figure 6.17 Double Lewis acid activation.

substitutionally inert; consequently, we have extended our studies with dinuclear Cu(II) complexes (**25** and **26**), which are substitutionally labile (Figure 6.17) [76–78]. X-ray crystallographic studies reveal that a phosphate can bridge the two metal centers in both **25** and **26**, as it does in **24**. The intermetal distances in **24**–**26** are about 2.9, 3.6 and 5.0 Å, respectively. Intermetal distances in dinuclear metalloenzymes that process phosphate esters also range between about 3 and 5 Å [1]. The dinuclear metal centers in **25** and **26** provide five- to six-orders of magnitude rate acceleration for the transesterification of 2-hydroxypropyl *p*-nitrophenyl phosphate and RNA. Based on the results of studies on cobalt complex **24** and crystal structures and kinetic studies of the copper complexes, we proposed that the above rate accelerations for the transesterification reaction are at least in part due to double Lewis acid activation.

The metal hydroxide in **26** may act as an intramolecular general base catalyst for cleaving RNA. It is unlikely that the metal hydroxide acts as an intramolecular nucleophilic catalyst since the 2′-hydroxyl group of RNA provides enormous rate enhancement for cleaving RNA. Base-catalyzed cleavage of RNA is about a billion times faster than base-catalyzed hydrolysis of DNA due to the well positioned nucleophilic group of RNA [53]. Double Lewis acid activation is expected to provide comparable rate acceleration for cleaving RNA and DNA. However, this mode of activation by itself will have a more significant effect on RNA cleavage. DNA would require nine additional orders of magnitude rate acceleration to be as reactive as RNA.

6.8
Phosphatase Models

In nature, many enzymes that hydrolyze phosphate monoesters are activated by two or more metal ions. They include alkaline phosphatase [79], purple acid phosphatase [80], inositol monophosphatase [81], and D-fructose 1,6-biphosphate 1-phosphatase [82]. The active sites of protein serine and threonine phosphatases also consist of dinuclear

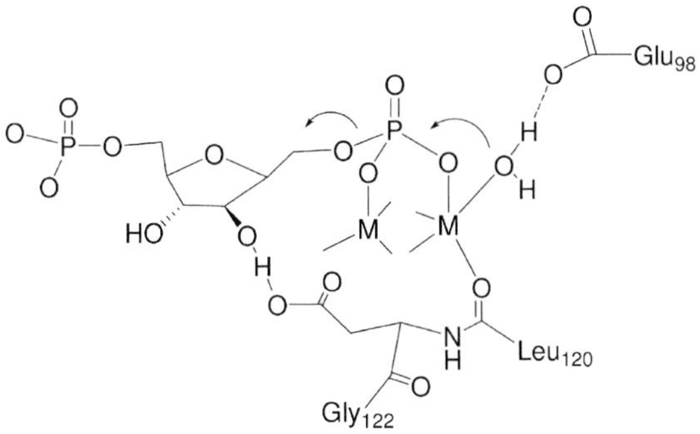

Figure 6.18 Active site of D-fructose 1,6-biphosphate 1-phosphatase.

metal centers [4–7, 83]. There has been considerable, prolonged interest in understanding the mechanism of hydrolysis of phosphate monoesters and the role of simple mononuclear [84–87] and dinuclear [88–90] metal complexes in the hydrolysis reaction. Since the hydrolysis of phosphate mono- and diesters often follows different mechanisms [19], it is important to study the mechanisms of both reactions. In the absence of added catalysts, monoesters and diesters generally hydrolyze by dissociative and associative mechanisms, respectively.

The structure of D-fructose 1,6-biphosphate 1-phosphatase has been reported [82]. Proposedly, the two metal centers at the active site of the enzyme are bridged by the substrate phosphate monoester, and a metal hydroxide is involved in the hydrolysis (Figure 6.18).

Sargeson et al. [91] showed that a phosphate monoester bridging two Co(III) centers is rapidly hydrolyzed by a neighboring metal hydroxide (27) (Figure 6.19). The metal-bound phosphate is hydrolyzed (4×10^{-1} s^{-1} at 25 °C) about eight orders of magnitude more rapidly than the free phosphate (2×10^{-9} s^{-1} at 25 °C). More recently, we showed that the phosphate monoester in 28 [92] is hydrolyzed (5×10^{-2} s^{-1} at 25 °C) about 11 orders of magnitude more rapidly than the free phosphate (5×10^{-13} s^{-1} at 25 °C) (Figure 6.19). Double Lewis acid activation may not provide as much rate acceleration for hydrolyzing phosphate monoesters as for hydrolyzing phosphate diesters (24). In

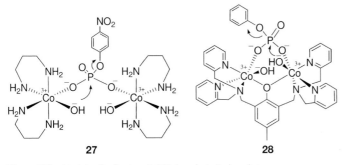

27   28

Figure 6.19 Models of D-fructose 1,6-biphosphate 1-phosphatase.

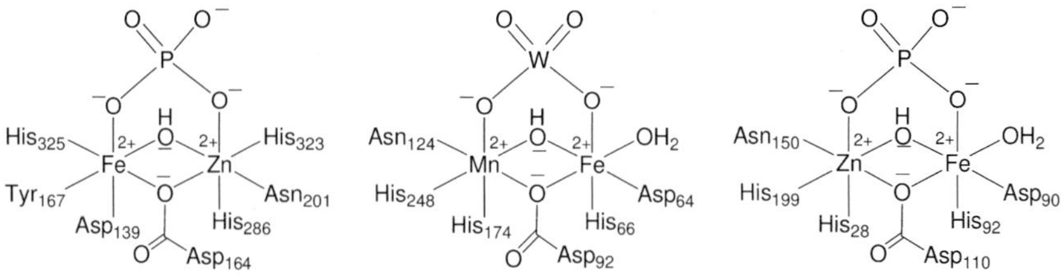

Figure 6.20 Active sites of phosphatases with a dimetallic diamond core.

phosphate monoester hydrolysis, the developing negative charge on the phosphoryl oxygen is not directly stabilized by the two metal ions. Thus, the metal hydroxide probably provides much of the observed rate acceleration.

The structures of several interesting phosphatases with a "dimetallic diamond core" have been determined (kidney bean purple acid phosphatase (KBPAP) [80], protein phosphatase-1 (PP-1) [4–7], and calcineurin α (PP-2B) [4–7]). Although the amino acid sequences of the three enzymes are unrelated, the active site structures are strikingly similar. These enzymes all catalyze the hydrolysis of phosphate monoesters and have a dimetallic diamond core. The active site of KBPAP contains a dinuclear $Zn(II)$–$Fe(II)$ center with an intermetal distance of 3.1 Å. Both metal ions are held in octahedral coordination geometries and are connected by two oxygen bridges, forming the four-membered-ring diamond core. Inorganic phosphate, the hydrolysis reaction product, bridges the two metal centers in the enzyme (**29**) (Figure 6.20). The active site structure of PP-1, with a tungstate inhibitor bound, also shows two metal ions [$Mn(II)$ and $Fe(II)$] held in very similar coordination spheres (**30**), with tungstate bridging the two metal ion centers. This enzyme acts on phosphorylated serine and threonine side chains, thereby playing an important role in signal transduction [93, 94]. PP-2B has a similar role, and its active site has a very similar ligand grouping around $Zn(II)$ and $Fe(III)$ ions (**31**).

A simple $Co(III)$ complex (**32**) whose structure closely resembles those of the above-mentioned enzymes (**29–31**) has been developed [95]. ^{18}O labeling studies showed that an oxide group that is bridging the two metal centers cleaves the bridging phosphate monoester in **32** (Figure 6.21). Kinetic isotope effect studies found that p-nitrophenyl

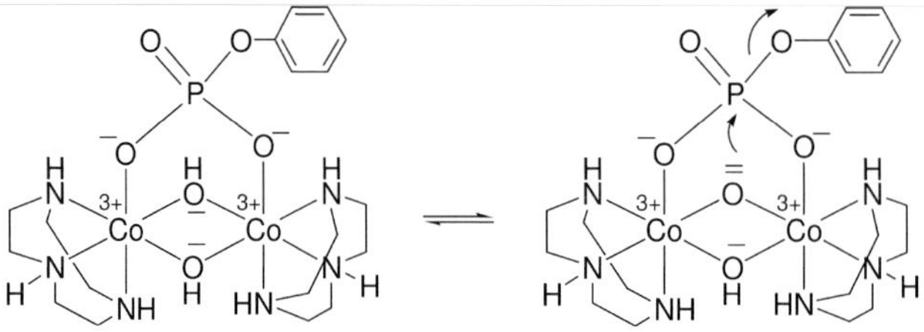

32

Figure 6.21 Phosphatase model with dimetallic a diamond core.

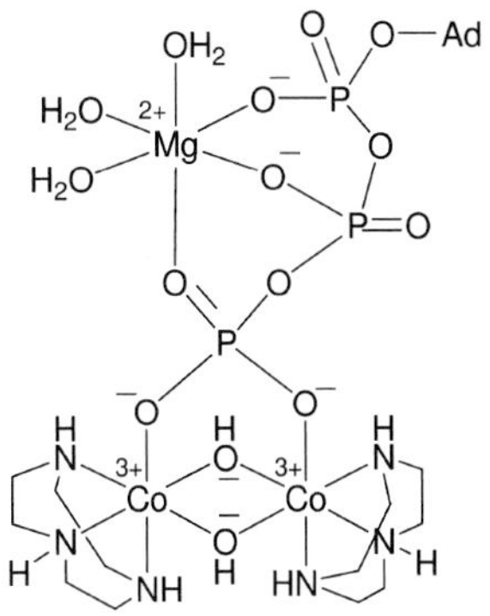

33

Figure 6.22 ATP hydrolysis with Co(iii) and Mg(ii).

phosphate bridging the two metal centers in **32** is hydrolyzed by a concerted mechanism with both bond making (nucleophile) and breaking (leaving group) processes at the transition state [96]. In contrast, the mechanism for hydrolysis of the free phosphate is largely dissociative in character, with very little bond formation to the nucleophile at the transition state.

It is interesting to compare the reactivities of **32** and **28**. Phenyl phosphate in **28** [92] is hydrolyzed about three orders of magnitude more rapidly than phenyl phosphate in **32** at pH 7. However, comparing the rate constants of the fully ionized species, **32** is about two orders of magnitude more reactive than **28**. In the hydrolysis of ATP to ADP and inorganic phosphate (**33**) [97], a third metal ion [Mg(ii)] can be used to give leaving-group activation in addition to the double Lewis acid and nucleophile activations provided by the other two metal ions [Co(iii)] (Figure 6.22). Remarkably, Mg(ii) provides a ca. 3000-fold rate acceleration for the hydrolysis.

6.9
Phosphodiesterase Models

There has been much interest in developing artificial metalloenzymes that hydrolyze the phosphate diester bonds of RNA and DNA. Chemical catalysts that hydrolyze nucleic acids may provide valuable insights into how nucleases work. In addition, catalysts that hydrolyze RNA sequence specifically [98–102] or DNA sequence specifically [103] may be useful as therapeutic agents. Indeed, sequence-specific cleavage of RNA is thought to be a natural defense mechanism against invading viruses [104]. Numerous transition metal [105–113] and lanthanide complexes [114–119] that hydrolyze phosphate diesters have been reported. Dinuclear Co(iii) [120, 72, 75], Cu(ii) [76, 77, 121–123] and Zn(ii) [124–133] complexes are even more reactive than their mononuclear analogs for hydrolyzing phosphate diesters. Not surprisingly, the active sites of many phosphodiesterases contain two metal ions [1–3].

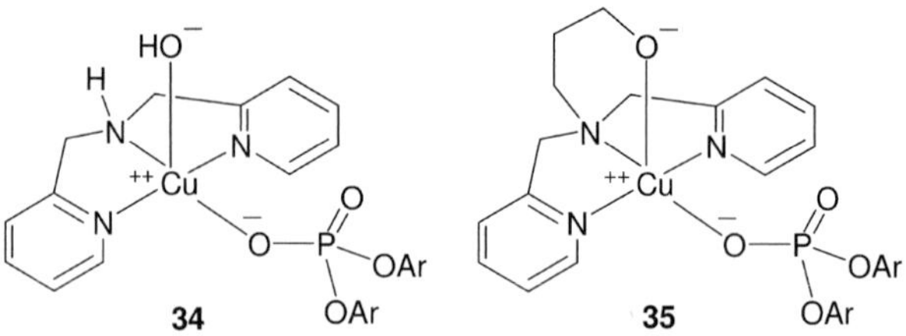

Figure 6.23 Metal hydroxide (**34**) and metal alkoxide (**35**) used for cleaving phosphate diesters.

Metal hydroxide, metal alkoxide and oxide bridging two metal centers are all effective nucleophiles for cleaving phosphate diesters. Metal hydroxides provide comparable rate accelerations for hydrolyzing phosphate diesters with good or poor leaving groups, yet metal alkoxides provide far greater rate acceleration for hydrolyzing phosphates with good leaving groups than for those with poor leaving groups. Thus, **6** provides a ca. 10 orders of magnitude rate acceleration for hydrolyzing bis-(p-nitrophenyl)phosphate [53] and dimethyl phosphate [43]. The metal alkoxide in **35** (Figure 6.23) is more reactive than the corresponding metal hydroxide (**34**) for cleaving phosphate diesters with very good leaving groups (2,4-dinitrophenol) but the order of the reactivity is reversed for cleaving phosphate diesters with poorer leaving groups (p-nitrophenol) [134].

Strongly basic nucleophiles rapidly cleave carboxyl esters with weakly basic leaving groups [52]. The rate of displacement of strongly basic leaving groups with weakly basic nucleophiles is very slow. Metal-alkoxides are too weakly basic to displace strongly basic leaving groups such as methoxide yet they are basic enough to rapidly displace weakly basic leaving groups like 2,4-dinitrophenolate. So why should metal hydroxides provide comparable rate acceleration for hydrolyzing phosphates with good or poor leaving groups? It appears that for cleaving phosphates with poor leaving groups, a deprotonatable nucleophile (such as a metal hydroxide but not a metal alkoxide) is required. The expulsion of poor leaving groups (as in RNA and DNA hydrolysis without leaving-group activation) is probably accompanied by deprotonation of the metal hydroxide (**36**) (Figure 6.24).

The phosphate diester in **37** (Figure 6.25) is cleaved about 11 orders of magnitude more rapidly than the uncoordinated diester [72]. The mechanism of the reaction involves nucleophilic attack of the bridging phosphate by the bridging oxide. Assuming

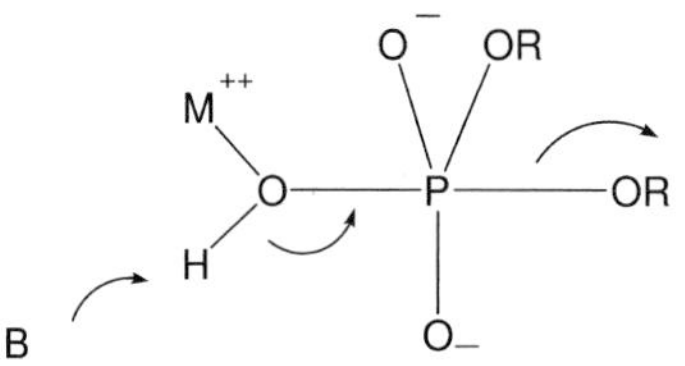

Figure 6.24 Deprotonatable metal hydroxide.

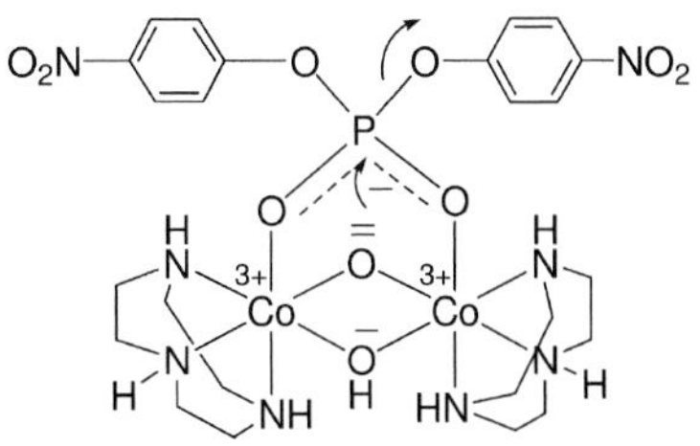

Figure 6.25 Bridging oxide and metal hydroxide.

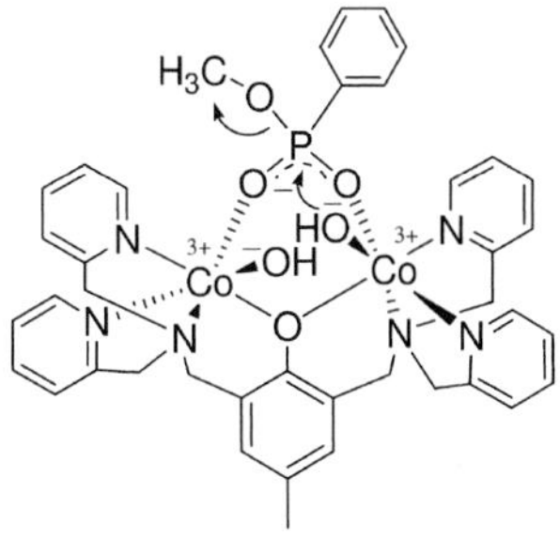

that double Lewis acid activation provides a ca. five orders of magnitude rate acceleration, the oxide should provide ca. six orders of magnitude rate acceleration for the hydrolysis. Although the dinuclear metal complex provides large rate accelerations for hydrolyzing phosphate diesters with good leaving groups, it is ineffective for hydrolyzing phosphates with poor leaving groups. In this respect, the bridging oxide behaves like the metal alkoxide (**35**). Neither nucleophile can be deprotonated like the metal hydroxide (**36**). In contrast to the dinuclear complex with the bridging oxide, the dinuclear complex with the metal hydroxide (**38**) (Figure 6.25) provides about 11 orders of magnitude rate acceleration for hydrolyzing phosphate diesters with good or poor leaving groups [120]. Thus, the effect of metal-hydroxide activation and leaving-group activation are additive. Since metal alkoxides and oxides bridging two metal centers provide far greater rate acceleration for hydrolyzing phosphates with good leaving groups than for hydrolyzing phosphates with poor leaving groups, synergistic effect between nucleophile and leaving-group activations is anticipated.

Methyl-*p*-nitrophenyl phosphate coordinated to the two metal centers in **37** undergoes hydrolysis by a two-step addition–elimination mechanism [73]. The free phosphate hydrolyzes by a concerted mechanism. In both phosphate monoester and diester hydrolysis, the two Co(III) centers in **32** and **37** stabilize the five-coordinate phosphate species (transition state or intermediate) by bringing the phosphate and nucleophile together. This stabilization leads to a change in mechanism from dissociative to concerted for a phosphate monoester hydrolysis [96] and from concerted to stepwise for phosphate diester hydrolysis [73].

6.10
Polymerases and DNases

All life forms require polymerases for replication of nucleic acids. DNA polymerases from various life forms [8] as well as reverse transcriptase from HIV [9] are all activated by two metal ions, which are thought to provide three types of activation, i.e. Lewis acid, metal-hydroxide and leaving group [Figure 6.26(A)]. The Klenow fragment of DNA polymerase I is a 3′,5′-exonuclease that is involved in editing the growing

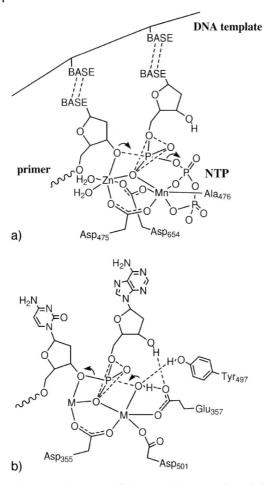

Figure 6.26 Active sites of (A) DNA polymerase and (B) 3′,5′-exonuclease.

DNA chain. Based on crystal data, Steitz and Beese [135] proposed an interesting mechanism for the dinuclear metalloenzyme catalyzed hydrolysis of DNA [Figure 6.26(B)]. The proposed mechanism for the breakdown of DNA is remarkably similar to that proposed for DNA synthesis, involving activations by metal hydroxide, leaving group and two Lewis acids [136]. In DNA synthesis, the leaving group is already a good one and the leaving oxygen does not need to be directly coordinated to the metal.

In model systems (**6**), up to 10 orders of magnitude rate acceleration has been obtained by combining Lewis acid and metal-hydroxide activations [43, 53]. Leaving-group activation is expected to be additive to Lewis acid and metal-hydroxide activations since **6** provides comparable rate acceleration for hydrolyzing phosphate diesters with good or poor leaving groups. If the metal lowers the basicity of the leaving group by 10 log units, the rate of hydrolysis should increase about ten million-fold since the

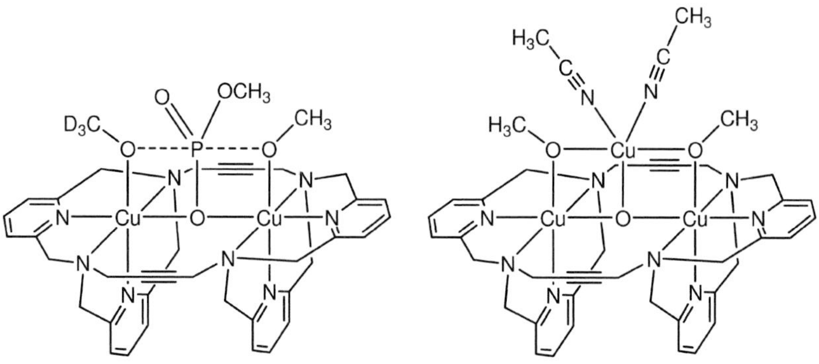

Figure 6.27 A 3',5'-exonuclease model.

Brønsted coefficient (β_{lg}) is around 0.7. Thus, Lewis acid activation, metal-hydroxide activation and leaving-group activation could add up to over 10^{17}-fold rate acceleration for DNA hydrolysis. The pseudo-first order rate constant for hydrolysis of a phosphate diester bond of DNA has been estimated to be about 10^{-19} s^{-1} at neutral pH and 25 °C [13]. A 10^{17}-fold rate acceleration for hydrolyzing DNA would reduce the half-life of the phosphate diester bond from billions of years to about a minute or two.

A remarkable model system (**39**) (Figure 6.27) that combines some of the most salient features of the polymerase and the exonuclease has been reported [137]. The dinuclear metal complex **39** is highly efficient for catalyzing the transesterification of dimethyl phosphate in methanol. This is an interesting result given the stability of dimethyl phosphate (Table 6.1). The proposed mechanism for the transesterification reaction involves two Lewis acid activations, metal-hydroxide activation and leaving-group activation just as in the proposed mechanism for the exonuclease [Figure 6.26(B)]. Interestingly, the crystal structure of a transition state analog (**40**) for the transesterification reaction has been obtained (Figure 6.27). Leaving-group activation is important for the transesterification reaction since metal alkoxides (**35**) cannot be used to displace rapidly strongly basic alkoxides.

6.11
Conclusion

In summary, metal complexes can provide large rate-accelerations for hydrolyzing esters, amides, nitriles, and phosphate mono-, di- and triesters. Combination of Lewis acid activation and a well-positioned metal hydroxide is all that is required for hydrolyzing carboxyl esters, amides and nitriles under mild conditions (neutral pH, ambient temperature). Dinuclear metal complexes are particularly effective for hydrolyzing amides and nitriles. Double Lewis acid activation alone is enough for efficient cleavage of RNA but not for hydrolyzing DNA. Combination of Lewis acid activation, nucleophile activation and leaving-group activation is required to hydrolyze the phosphate diester bonds of DNA under mild conditions. Over the past several decades, much has been learned about the mechanistic role of metal ions for a wide range of hydrolysis reactions. Model studies with mono and dinuclear metal complexes have allowed us to better understand the modes of activations that metal ions can provide and the cooperativity of metal ions in the hydrolysis reactions. Although much progress has been made over the past decade, developing evermore-stable yet reactive metal complexes for the sequence-specific hydrolysis of proteins and nucleic acids remains an important challenge. Stereoselective hydrolysis of amides and esters and obtaining maximum catalytic turnover for the hydrolysis reactions are some of the other remaining challenges.

References

1 N. Sträter, W. N. Lipscomb, T. Klabunde, B. Krebs, *Angew. Chem. Int. Ed. Engl.* **1996**, *35*, 2024–2055.

2 D. E. Wilcox, *Chem. Rev.* **1996**, *96*, 2435–2458.

3 W. N. Lipscomb, N. Sträter, *Chem. Rev.* **1996**, *96*, 2375–2433.

4 J. Goldberg, H.-B. Huang, Y.-G. Kwon, P. Greenbard, A. C. Nairn, J. Kuriyan, *Nature* **1995**, *376*, 745.

5 M.-P. Egloff, P. T. W. Cohen, P. Reinemer, D. Barford, *J. Mol. Biol.* **1995**, *254*, 942.

6 J. P. Griffith, J. L. Kim, E. E. Kim, M. D. Sintchak, J. A. Thomson, M. J. Fitzgibbon, M. A. Fleming, P. R. Caron, K. Hsaio, M. A. Navia, *Cell* **1995**, *82*, 507.

7 C. R. Kissinger, H. E. Parge, D. R. Knighton, C. T. Lewis, L. A. Pelletier, A. Tempczyk, V. J. Kalish, K. D. Tucker, R. E. Showalter, E. W. Moomaw, L. N. Gastinel, N. Habucka, X. Chen, F. Maldonado, J. E. Barker, R. Bacquet, E. Villafranca, *Nature* **1995**, *378*, 641.

8 S. Doublie, S. Tabor, A. M. Long, C. C. Richardson, T. Ellenberger, *Nature* **1998**, *391*, 251–258.

9 J. F. Davis II, Z. Hostomska, Z. Hostomsky, S. R. Jordan, D. A. Matthews, *Science* **1991**, *252*, 88–95.

10 J. A. Doudna, T. R. Cech, *Nature* **2002**, *418*, 222–228.

11 J. Suh, *Acc. Chem. Res.* **2003**, *36*, 562–570.

12 J. Suh, *Acc. Chem. Res.* **1992**, *25*, 273–279.

13 N. H. Williams, B. Takasaki, M. Wall, J. Chin, *Acc. Chem. Res.* **1999**, *32*, 485–493.

14 J. Chin, *Acc. Chem. Res.* **1991**, *24*, 145–152.

15 R. Breslow, *Acc. Chem. Res.* **1995**, *28*, 146–153.

16 J. P. Guthrie, P. A. Cullimore, *Can. J. Chem.* **1980**, *58*, 1281–1294.

17 T. Yamana, Y. Mizukami, A. Tsuji, Y. Yasuda, K. Masuda, *Chem. Pharm. Bull.* **1970**, *20*, 881.

18 R. M. Smith, D. E. Hansen, *J. Am. Chem. Soc.* **1998**, *120*, 8910–8913.

19 J. P. Guthrie, *J. Am. Chem. Soc.* **1977**, *99*, 3991–4001.

20 J. P. Guthrie, *J. Am. Chem. Soc.* **1973**, *95*, 6999–7003.

21 T. P. Yoon, E. N. Jacobsen, *Science* **2003**, *299*, 1691–1693.

22 J. S. Johnson, D. A. Evans, *Acc. Chem. Res.* **2000**, *33*, 325–335.

23 M. Santelli, J.-M. Pons **1996**, *Lewis Acids and Selectivity in Organic Synthesis*, CRC Press, New York.

24 A. Pfaltz, *Acc. Chem. Res.* **1993**, *26*, 339–345.

25 K. A. Jorgensen, M. Johannsen, S. Yao, H. Audrain, J. Thorhauge, *Acc. Chem. Res.* **1999**, *32*, 605–613.

26 T. Cech, B. L. Bass, *Annu. Rev. Biochem.* **1986**, *55*, 599–629.

27 S. Linskog, A. Liljas, *Curr. Opin. Struct. Biol.* **1993**, *3*, 915–920.

28 Y. Pocker, J. T. Stone, *J. Am. Chem. Soc.* **1965**, *87*, 5497–5498.

29 Y. Pocker, S. Sarkanen, *Adv. Enzymol.* **1987**, *47*, 149.

30 P. Woolley, *Nature* **1975**, *258*, 677.

31 P. Woolley, *J. Chem. Soc., Perkin Trans. 2* **1977**, 318.

32 E. Kimura, T. Shiota, T. Koike, M. Shiro, M. Kodama, *J. Am. Chem. Soc.* **1990**, *112*, 5805–5811.

33 W. P. Jencks **1969**, *Catalysis in Chemistry and Enzymology*, McGraw-Hill, New York, 175.

34 J. R. Murdoch, *J. Am. Chem. Soc.* **1972**, *94*, 4410–4418.

35 B. K. Takasaki, J. H. Kim, E. Rubin, J. Chin, *J. Am.* Chem. *Soc.* **1993**, *115*, 1157–1159.

36 D. A. Buckingham, J. M. Harrowfield, A. M. Sargeson, *J. Am. Chem. Soc.* **1974**, *96*, 1726–1729.

37 N. J. Curtis, A. M. Sargeson, *J. Am. Chem. Soc.* **1984**, *106*, 625–630.

38 D. A. Buckingham, F. R. Keene, A. M. Sargeson, *J. Am. Chem. Soc.* **1973**, *95*, 5649–5652.

39 P. Hendry, A. M. Sargeson, *J. Chem. Soc., Chem. Commun.* **1984**, 164.

40 E. Chaffee, T. P. Dasgupta, G. M. Harris *J. Am. Chem. Soc.* **1973**, *95*, 4169–4173.

41 J. H. Kim, J. Britten, J. Chin, *J. Am. Chem. Soc.* **1993**, *115*, 3618–3622.

42 A. J. Gordon, R. A. Ford **1972**, *The Chemist's Companion*, John Wiley & Sons: New York.

43 J. H. Kim, J. Chin, *J. Am. Chem. Soc.* **1992**, *114*, 9792–9795.

44 W. P. Jencks, *Adv. Enzymol.* **1975**, *43*, 219–240.

45 A. J. Kirby, *Adv. Phys. Org. Chem.* **1980**, *17*, 183.

46 R. Breslow, R. Fairweather, J. Keana, *J. Am. Chem. Soc.* **1967**, *89*, 2135–2138.

47 D. A. Buckingham, P. Morris, A. M. Sargeson, A. Zanella, *Inorg. Chem.* **1977**, *16*, 1910–1923.

48 D. A. Buckingham, F. R. Keene, A. M. Sargeson, *J. Am. Chem. Soc.* **1974**, *96*, 4981–4983.

49 J. T. Groves, R. R. Chambers, Jr. *J. Am. Chem. Soc.* **1984**, *106*, 630–638.

50 J. Chin, X. Zou, *J. Am. Chem. Soc.* **1984**, *106*, 3687–3688.

51 J. Chin, V. Jubian, *J. Chem. Soc., Chem. Commun.* **1989**, 839.

52 W. P. Jencks, M. Gilchrist, *J. Am. Chem. Soc.* **1967**, *90*, 2622–2637.

53 J. Chin, B. Banaszczyk, V. Jubian, X. Zou, *J. Am. Chem. Soc.* **1989**, *111*, 186–190.

54 A. M. Davis, A. D. Hall, A. Williams, *J. Am. Chem. Soc.* **1988**, *110*, 5105–5108.

55 P. Hendry, A. M. Sargeson, *Inorg. Chem.* **1990**, *29*, 92.

56 J. Chin, M. Banaszczyk, *J. Am. Chem. Soc.* **1989**, *111*, 2724–2726.

57 D. W. Christianson, W. N. Lipscomb, *Acc. Chem. Res.* **1989**, *22*, 62–69.

58 N. J. Curtis, K. S. Hagen, A. M. Sargeson, *J. Chem. Soc., Chem. Commun.* **1984**, 1571–1573.

59 N. N. Murthy, M. Mahroof-Tahir, K. D. Karlin, *J. Am. Chem. Soc.* **1993**, *115*, 10404–10405.

60 C. F. Meares, T. M. Rana, *J. Am. Chem. Soc.* **1991**, *113*, 1859–1861.

61 A. Schepartz, B. Cuenoud, *J. Am. Chem. Soc.* **1990**, *112*, 3247–3249.

62 D. Hoyer, H. Cho, P. G. Schultz, *J. Am. Chem. Soc.* **1990**, *112*, 3249–3250.

63 N. M. Miovic, J. D. Badjic, N. M. Kostic, *J. Am. Chem. Soc.* **2004**, *126*, 696–697.

64 C. E. Yoo, P. S. Chae, J. E. Kim, E. J. Jeong, J. Suh, *J. Am. Chem. Soc.* **2003**, *125*, 14 580–14 589.

65 M. Tokunaga, J. F. Larrow, F. Kakiuchi, E. N. Jacobsen, *Science* **1997**, *277*, 936–938.

66 G. M. Sammis, H. Danjo, E. N. Jacobsen, *J. Am. Chem. Soc.* **2004**, *126*, 9928–9929.

67 B. M. Trost, V. S. C. Yeh, H. Ito, N. Bremeyer, *Org. Lett.* **2002**, *4*, 2621–2623.

68 B. M. Trost, A. Fettes, B. T. Shireman, *J. Am. Chem. Soc.* **2004**, *126*, 2660–2661.

69 E. J. Corey, R. K. Bakshi, S. Shibata, C. P. Chen, V. K. Singh, *J. Am. Chem. Soc.* **1987**, *109*, 7925–7926.

70 P. A. Sutton, D. A. Buckingham, *Acc. Chem. Res.* **1987**, *20*, 357–364.

71 N. H. Williams, J. Chin, *J. Chem. Soc., Chem. Commun.* **1996**, 131.

72 N. H. Williams, W. Cheung, J. Chin, *J. Am. Chem. Soc.* **1998**, *120*, 8079–8087.

73 T. Humphry, M. Forconi, N. H. Williams, A. Hengge, *J. Am. Chem. Soc.* **2002**, *124*, 14 860–14 861.

74 D. A. Usher, D. I. Richardson, D. G. Oakenfull, *J. Am. Chem. Soc.* **1970**, *92*, 4699–4712.

75 D. Wahnon, A.-M. Lebuis, J. Chin, *Angew. Chem., Int. Ed. Engl.* **1995**, *34*, 2412.

76 M. Wall, R. C. Hynes, J. Chin, *Angew. Chem., Int. Ed. Engl.* **1993**, *32*, 1633.

77 M. J. Young, J. Chin, *J. Am. Chem. Soc.* **1995**, *117*, 10577–10578.

78 M. J. Young, Ph.D. Thesis, McGill University, Canada, 1996

79 E. E. Kim, H. W. Wyckoff, *J. Mol. Biol.* **1991**, *218*, 449.

80 K. Klabunde, N. Sträter, P. Tucker, H. Witzel, B. Krebs, *Science* **1995**, *268*, 1489.

81 S. J. Pollack, J. R. Atack, M. R. Knowles, G. McAllister, C. I. Ragan, R. Baker, S. R. Fletcher, L. L. Iversen, H. B. Broughton, *Proc. Natl. Acad. Sci. U.S.A.* **1994**, *91*, 5766.

82 Y. Zhang, J.-Y. Liang, S. Huang, H. Ke, W. Lipscomb, *Biochemistry* **1993**, *32*, 1844.

83 J. B. Vincent, B. A. Averill, *FEBS Lett.* **1990**, *263*, 265.

84 T. H. Fife, M. P. Pujari, *J. Am. Chem. Soc.* **1988**, *110*, 7790–7797.

85 D. R. Jones, L. F. Lindoy, A. M. Sargeson, *J. Am. Chem. Soc.* **1983**, *105*, 7327–7336.

86 J. M. Harrowfield, D. R. Jones, L. F. Lindoy, A. M. Sargeson, *J. Am. Chem. Soc.* **1980**, *102*, 7733–7741.

87 P. Hendry, A. M. Sargeson, *Prog. Inorg. Chem.* **1990**, *38*, 201.

88 D. R. Jones, L. F. Lindoy, A. M. Sargeson, *J. Am. Chem. Soc.* **1984**, *106*, 7807–7819.

89 D. H. Vance, A. W. Czarnik, *J. Am. Chem. Soc.* **1993**, *115*, 12 165–12 166.

90 J. Chin, M. Banaszczyk, *J. Am. Chem. Soc.* **1989**, *111*, 4103–4105.

91 P. Hendry, A. M. Sargeson, *Inorg. Chem.* **1990**, *29*, 92.

92 J. S. Seo, N.-D. Sung, R. C. Hynes, J. Chin, *Inorg. Chem.* **1996**, *35*, 7472.

93 D. Barford, *Trends Biochem. Sci.* **1996**, *21*, 407.

94 T. S. Widlanski, J. K. Myers, B. Stec, K. M. Holtz, E. R. Kantowitz, *Chem. Biol.* **1997**, *4*, 489–492.

95 N. H. Williams, A.-M. Lebuis, J. Chin, *J. Am. Chem. Soc.* **1999**, *121*, 3341–3348.

96 T. Humphry, M. Forconi, N. H. Williams, A. C. Hengge, *J. Am. Chem. Soc.* **2004**, *126*, 11864–11869.

97 N. H. Williams, *J. Am. Chem. Soc.* **2000**, *122*, 12 023–12 024.

98 K. Matsumura, M. Endo, M. Komiyama, *J. Chem. Soc., Chem. Commun.* **1994**, 2019–2020.

99 D. Magda, R. A. Miller, J. Sessler, B. L. Iverson, *J. Am. Chem. Soc.* **1994**, *116*, 7439–7440.

100 J. Hall, D. Huesken, U. Pieles, H. E. Moser, R. Haener, *Chem. Biol.* **1994**, 185–190.

101 J. K. Baskin, E. I. Frolova, U. Sampath, *J. Am. Chem. Soc.* **1994**, *116*, 5981–5982.

102 B. F. Baker, S. S. Lot, J. Kringler, S. Cheng-Flournoy, P. Viliet, H. M. Sasmor, A. M. Siwkowski, L. L. Chappell, J. R. Morrow, *Nucleic Acids Res.* **1999**, *17*, 1547–1551.

103 M. Komiyama **2003**, *The Lanthanides and Their Interrelations With Biosystems*, ed. H. Sigel, Marcel Dekker, Inc., New York, Vol. 40, 463–475.

104 G. Hutvgner, P. A. Zamore, *Science*, **2002**, *297*, 2056.

105 B. N. Trawick, A. T. Daniher, J. K. Bashkin, *Chem. Rev.* **1998**, *98*, 939.

106 D. M. Perreault, E. V. Anslyn, *Angew. Chem. Int. Ed.* **1997**, *36*, 432–450.

107 B. Linkletter, J. Chin, *Angew. Chem., Int. Ed. Engl.* **1995**, *34*, 472.

108 R. Breslow, D. L. Huang, *Proc. Natl. Acad. Sci. U.S.A.* **1991**, *88*, 4080.

109 R. Breslow, B. Zhang, *J. Am. Chem. Soc.* **1994**, *116*, 7893.

110 E. L. Hegg, K. A. Deal, L. Kiessling, J. N. Burstyn, *Inorg. Chem.* **1997**, *36*, 1715–1518.

111 R. Breslow, D. Berger, D.-L. Huang, *J. Am. Chem. Soc.* **1990**, *112*, 3686–3687.

112 H. Ait-Haddou, J. Sumaoka, S. L. Wiskur, J. F. Folmer-Anderson, E. V. Anslyn, *Angew. Chem. Int. Ed.* **2002**, *41*, 4014–4016.

113 L. Bonf, M. Gatos, F. Mancin, P. Tecilla, U. Tonellato, *Inorg. Chem.* **2003**, *42*, 3943–3949.

114 H.-J. Schneider, A. K. Yatsimirsky **2003**, *The Lanthanides and Their Interrelations With Biosystems*, ed. H. Sigel, Marcel Dekker, Inc., New York, Vol. 40, 369–462.

115 B. K. Takasaki, J. Chin, *J. Am. Chem. Soc.* **1995**, *117*, 8582.

116 P. Hurst, B. K. Takasaki, J. Chin, *J. Am. Chem. Soc.* **1996**, *118*, 9982.

117 B. K. Takasaki, J. Chin, *J. Am. Chem. Soc.* **1994**, *116*, 1121.

118 M. Yashiro, A. Ishiubo, M. Komiyama *Biochemistry* **1996**, *120*, 1067–1069.

119 J. R. Morrow, L. A. Buttrey, V. M. Shelton, K. A. Berback, *J. Am. Chem. Soc.* **1992**, *114*, 1903–1905.

120 J. S. Seo, R. C. Hynes, D. W. Williams, J. Chin, N.-D. Sung, *J. Am. Chem. Soc.*, **1998**, *120*, 9943–9944.

121 P. Molenveld, W. M. G. Stikvoort, H. Kooijman, A. L. Spek, J. F. J. Engbersen, D. N. Reinhoudt, *J. Am. Chem. Soc.* **1998**, *120*, 6726–6737.

122 S. Liu, A. D. Hamilton, *Chem. Commun.* **1999**, 587–588.

123 T. Gadja, A. Jancs, S. Mikkola, H. Lönnerg, H. Sirges, *J. Chem. Soc., Dalton Trans.* **2002**, 1757–1763.

124 W. H. Chapman, Jr., R. Breslow, *J. Am. Chem. Soc.* **1995**, *117*, 5462–5469.

125 M. Yashiro, A. Ishikubo, M. Komiyama, *J. Chem. Soc., Chem. Commun.* **1995**, 1793–1794.

126 P. Molenveld, W. M. G. Stikvoort, H. Kooijman, A. L. Spek, J. F. J. Engbersen, D. N. Reinhoudt, *J. Org. Chem.* **1999**, *64*, 3896–3906.

127 P. Rossi, F. Felluga, P. Tecilla, F. Formaggio, M. Crisma, C. Toniolo, P. Scrimin, *Biopolymers* **2000**, *55*, 496–501.

128 C. He, S. J. Lippard, *J. Am. Chem. Soc.* **2000**, *122*, 184–185.

129 S. Albedyhl, D. Schnieder, A. Jancs, T. Gadja, B. Krebs, *Eur. J. Inorg. Chem.* **2002**, 1400–1409.

130 K. Worm, F. Chu, K. Matsumoto, M. D. Best, V. Lynch, E. V. Anslyn, *Chem.-Eur. J.* **2003**, *9*, 741–747.

131 T. Gajda, R. Krämer, A. Jancs, *Eur. J. Inorg. Chem.* **2000**, 1635–1644.

132 K. Yamada, Y. Takahashi, H. Yamamura, S. Araki, K. Saito, M. Kawai, *Chem. Commun.* **2000**, 1315–1316.

133 S. Kawahara, T. Uchimaru, *Eur. J. Inorg. Chem.* **2001**, 2437–2442.

134 M. J. Young, D. Wahnon, R. C. Hynes, J. Chin, *J. Am. Chem. Soc.* **1995**, *117*, 9441–9447.

135 L. S. Beese, T. A. Steitz, *EMBO J.*, **1991**, *10*, 25.

136 T. A. Steitz, *Nature*, **1998**, *391*, 231.

137 U. Kühn, S. Warzeska, H. Pritzkow, R. Krämer, *J. Am. Chem. Soc.* **2001**, *123*, 8125–8126.

7
Artificial Restriction Enzymes As Tools For Future Molecular Biology and Biotechnology

Yoji Yamamoto and Makoto Komiyama

7.1
Introduction

Today, molecular biology and biotechnology play essential roles in science and industry. These technologies are very simple, but are so powerful that even high-school students can prepare rather complicated proteins in large amounts without much difficulty (of course, appropriate guidance should be provided). Furthermore, unexpected discoveries are being made worldwide almost everyday. Information accumulated by these studies enables medical and medicinal fields to be more successful and fruitful. These trends should increase over the next several decades. Unfortunately, almost all the materials used to date (e.g., enzymes, DNAs, vectors, and cells) are gifts from Mother Nature, and the contribution of chemists to this technology has not been very large. However, remarkable developments in bioorganic chemistry and supramolecular chemistry have marked the beginnings of a new epoch.

This chapter describes artificial enzymes that can hydrolyze DNA at the target-site with desired site-specificity [1]. With these totally man-made tools (artificial restriction enzymes), even huge DNA can be cut at a predetermined position and used for the preparation of recombinant DNA. This manipulation is hardly accomplishable as long as only naturally occurring restriction enzymes are used. Hopefully, these tools will open the way to new molecular biology and biotechnology in which the DNA of higher animals and plants can be manipulated.

7.2
Significance of Artificial Restriction Enzymes

Current molecular biology and biotechnology are based on two key processes: (i) site-selective scission of DNA by restriction enzymes and (ii) recombination of DNA fragments by another enzyme, known as ligase. These methods, first reported in the 1970s, are almost perfect as long as we manipulate small DNA such as the plasmid DNA of bacteria and genome DNA of virus, which are composed of several thousand

Artificial Enzymes. Edited by R. Breslow
Copyright © 2005 WILEY-VCH Verlag GmbH & Co. KGaA, Weinheim
ISBN: 3-527-31165-3

nucleotide base-pairs. Soon, however, we will certainly want to manipulate DNAs of higher animals and plants, and widen the scope of technological applications. Can we directly use the present methods there? The answer is absolutely "no", as clearly shown by the following simple statistical calculation. Naturally occurring restriction enzymes have rather poor site-specificity, although the selectivity is almost 100%. Most of them recognize a specific sequence composed of 4 or 6 DNA-bases and cut the DNA there. On average, the scission site of these restriction enzymes should come out at every 4^4 (256) or 4^6 (4096) DNA-bases (assuming that four kinds of nucleotides, A, G, C, and T, are randomly distributed in the DNA). These site-specificities are sufficient for site-selective scission of small DNA (e.g., plasmid DNA), but apparently too low to cut the huge DNAs of higher life forms at the desired site. For example, the DNA of human beings is composed of more than 10^9 DNA-bases. If this DNA is treated with a 6-base recognizing restriction enzyme, the scission should occur at more than 10^5 sites. Precise gene manipulation is impossible. Thus, artificial restriction enzymes, which have much higher site-specificity and can selectively hydrolyze huge DNA at a predetermined position, are crucially important for further developments. To cut only one-site in human DNA, for example, we have to recognize a 16- or longer base sequence ($4^{16} > 10^9$).

The importance of artificial restriction enzymes has been well understood by many chemists, and several challenging attempts have been made already. Although many difficulties remain, elegant work by many people is paving the way. This chapter deals mainly with the recent work of our laboratory on site-selective DNA scission and its applications to biotechnology.

7.3
Non-enzymatic Catalysts for DNA Hydrolysis

One of the most crucial obstacles for the preparation of artificial restriction enzymes is the enormous stability of phosphodiester linkages in DNA. Nucleophilic attack of OH^- towards phosphodiester linkage in DNA for its hydrolytic scission is prevented by electrostatic repulsion between these two negatively charged species (the half-life of one linkage for the hydrolytic scission at pH 7 and 25 °C is estimated to be 200 million years [2]). For a long time, DNA could not be hydrolyzed without using naturally occurring nucleases. However, superb catalysts have been developed and this obstacle has been gradually overcome. In the 1990s, it was found that rare earth metal ions are remarkably active for the hydrolysis of phosphoesters and efficiently hydrolyze DNA at reasonable rates under physiological conditions [3] (rare earth metal ions are also highly active for the hydrolysis of RNA [4]). For DNA hydrolysis, the Ce(IV) ion is especially active. For example, 10 mM Ce(IV) decreases the half-life of dinucleotides to only several hours [5]. Compared with their uncatalyzed hydrolysis, an almost 10^{12}-fold acceleration has been achieved. When oligo- or polynucleotides are hydrolyzed, scission takes place almost randomly without specific-base preference. Although Ce(IV) is a well-known oxidant, no concurrent oxidative scission of DNA occurs, and the DNA scission proceeds totally via a hydrolytic pathway [5, 6]. This is a significant advantage of Ce(IV), since the DNA fragments obtained by the scission can be, when necessary, recombined with other fragments by using ligase (this enzyme is effective for the recombination of hydrolytic scission products but cannot recombine the products of oxidative scission). The second key step in molecular biology and biotechnology is successfully achievable. Thus, Ce(IV) and its complexes are satisfactorily compatible with the current technology, and are very promising for the catalytic center of artificial restriction enzymes. Other tetravalent metal ions, Th(IV) [7] and Zr(IV) [8], are also active for DNA hydrolysis. However, their activities are considerably smaller than that of Ce(IV). Even today, Ce(IV) is still the sole catalyst that can promptly hydrolyze linear DNAs irrespective of DNA sequences.

7.4
Molecular Design of Artificial Restriction Enzymes (Covalent vs. Non-Covalent Strategy)

7.4.1
Covalent Strategy for the First-generation of Artificial Restriction Enzymes

Artificial restriction enzymes were first prepared by conjugating "molecular DNA scissors" with a "sequence-recognizing moiety". The "molecular DNA scissors" are concentrated near the target phosphodiester linkages, and preferentially hydrolyze them because of favorable activation-entropy term. For example, the Ce(IV) ion was combined with an oligonucleotide (sequence-recognizing moiety) that is complementary with the substrate DNA near the target site and bears an iminodiacetate group at its 5'-end. The substrate DNA was selectively hydrolyzed at the target-site [9] [Figure 7.1(a)]. Similarly, tris(hydroxymethyl)aminomethane-modified peptide nucleic acid (PNA: sequence-recognizing moiety) was combined with Zr(IV) ion as molecular DNA scissors [10].

This "covalent" strategy is very successful for site-selective DNA scission. However, the scission is not sufficiently efficient and must be promoted for practical applica-

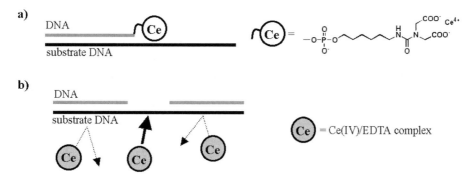

Figure 7.1 Site-selective hydrolysis of DNA using (a) covalent and (b) non-covalent strategies.

tions. The primary reason for the suppressed efficiency is that the catalytic activity of metal ion is notably diminished upon the conjugation with sequence-recognizing moieties. This effect is more drastic when strong ligands are used to bind the Ce(IV) tightly and place it near the target site. With weaker ligands, however, the metal ion is fixed less strictly and the site-selectivity tends to become lower.

7.4.2
Non-covalent Strategy for the Second-generation of Artificial Restriction Enzymes

To solve the above dilemma, a "non-covalent" strategy has been proposed [11]. Here, target phosphodiester linkages are differentiated from other linkages in terms of intrinsic reactivity. DNA scission should be restricted to the target site simply because the target phosphodiester linkages are more reactive than the others. Accordingly, covalent fixation of Ce(IV) to sequence-recognizing oligonucleotides is not necessary, and thus the intrinsic catalytic activity of this metal ion is fully utilized. As shown below, this "non-covalent" strategy is very successful in providing useful tools for future molecular biology and biotechnology.

7.4.3
Chemical Basis for "Non-covalent" Strategy

To accomplish site-selective DNA scission using the "non-covalent" strategy, we need both (i) molecular scissors that show sufficiently high substrate-specificity and (ii) "hot spots" that are formed at predetermined positions in substrate DNA and are hydrolyzed preferentially by these molecular scissors.

7.4.3.1 Molecular Scissors showing High Substrate-specificity

Although Ce(IV) ion is very active for DNA hydrolysis, it easily forms a metal hydroxide gel at physiological pH, and the system becomes heterogeneous. This imposes significant limitations to its practical applications. However, the Ce(IV)/EDTA complex is homogeneous under physiological conditions and efficiently hydrolyzes DNA (EDTA = ethylenediamine-N,N,N',N'-tetraacetic acid) [12]. Homogeneous solutions of this complex can be prepared simply by mixing stoichiometric amounts of $Ce(NH_4)_2(NO_3)_6$ and EDTA (4Na salt) in buffer solutions.

Notably, the Ce(IV)/EDTA complex shows remarkable substrate specificity. It effectively hydrolyzes polynucleotides and oligonucleotides that are longer than tetranucleotides. However, neither dinucleotides nor trinucleotides are hydrolyzed. Because of this characteristic, this complex had long been believed to be inactive for DNA hydrolysis. In addition to this specificity with respect to substrate size, this complex clearly differentiates between single- and double-stranded DNA. Thus, single-stranded DNA is efficiently hydrolyzed, but double-stranded DNA is hardly cleaved [13].

7.4.3.2 Formation of "Hot Spots" for Hydrolytic Scission at a Predetermined Position

These findings can be extended to "non-covalent" site-selective DNA hydrolysis [13]. With single-stranded DNA as substrate, the additives used are two oligonucleotides that are complementary with part of the substrate DNA. Upon mixing both these additives with substrate DNA, most of the DNA forms duplexes with the oligonucleotide additives and only the target site is kept single-stranded (this kind of structure is called as a gap). The position of the gap and its size (number of unpaired nucleotides) depend of course on the oligonucleotide additives and, thus, are easily controllable. These gap-sites are "hot spots" for catalysis by Ce(IV)/EDTA and are selectively hydrolyzed [Figure 7.1(b)]. Conversely, double-stranded portions in the substrate DNA are kept intact because of the poor reactivity. Covalent immobilization of the Ce(IV) complex to the target site is unnecessary.

Alternative "hot spots" are bulge-sites in substrate DNA, formed by using an oligonucleotide additive that is several nucleotides shorter than the DNA substrate [13]. To make "hot spots" in double-stranded DNA substrates, invasion of peptide nucleic acid (PNA) is quite useful, as detailed in Section 7.6.

7.5
Site-selective Scission of Single-stranded DNA

7.5.1
Promotion of Gap-selective DNA Hydrolysis by Introducing Monophosphate Groups to the Gap-site

As described above [Figure 7.1(b)], gap-sites in single-stranded DNA substrate, formed by "unmodified" oligonucleotides, can be selectively hydrolyzed by Ce(IV)/EDTA complex. However, both scission efficiency and site-selectivity are not satisfactorily high for practical applications. To promote these two factors, monophosphate groups are introduced to both edges (or either of them) of the gap-site [14]. Both scission efficiency and site-selectivity are remarkably improved. The monophosphate groups bind the Ce(IV) complex and recruit it to the target site (Figure 7.2).

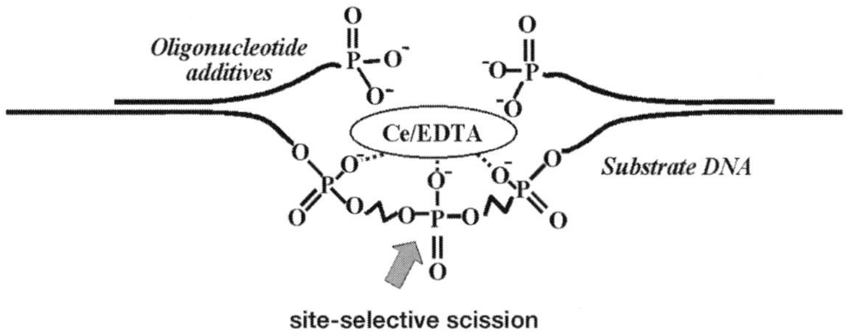

Figure 7.2 Promotion of gap-selective DNA hydrolysis by introducing monophosphate groups to the gap-site.

7.5.1.1 Selectivity and Efficiency of Site-selective DNA Scission

In the oligonucleotide additives used, a monophosphate group is bound to either the 3'- or the 5'-end via a $(CH_2)_{12}$ linker (Figure 7.3). When these additives are combined with the substrate DNA, gap-structures are formed at a predetermined site in the

Substrate DNA

DNA$^{(S5)}$: 5'-CAATTAGAATCAGGAATGGC**TTATG**GTGCAGACTGTCGACCTAAG-3'
DNA$^{(S3)}$: 5'-CAATTAGAATCAGGAATGGC**TTA**GTGCAGACTGTCGACCTAAG-3'
DNA$^{(S2)}$: 5'-CAATTAGAATCAGGAATGGC**TT**GTGCAGACTGTCGACCTAAG-3'
DNA$^{(S1)}$: 5'-CAATTAGAATCAGGAATGGC**T**GTGCAGACTGTCGACCTAAG-3'

Gap-forming Additives

DNA$^{(L)}$-L$_{12}$-P: 3'-GTTAATCTTAGTCCTTACCG-L$_{12}$-P-5'
P-L$_{12}$-DNA$^{(R)}$:   3'-P-L$_{12}$-CACGTCTGACAGCTGGATTC-5'
DNA$^{(L)}$-H: 3'-GTTAATCTTAGTCCTTACCG-5'
H-DNA$^{(R)}$:   3'-CACGTCTGACAGCTGGATTC-5'

Figure 7.3 DNA substrates and oligonucleotide additives used for site-selective scission of single-stranded DNA.

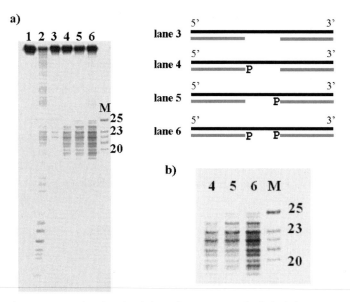

Figure 7.4 (a) Polyacrylamide gel electrophoresis patterns for the hydrolysis of DNA$^{(S5)}$ (^{32}P-labelled at the 5'-end) by Ce(ɪᴠ)/EDTA complex in the presence of various additive oligonucleotides at pH 7.0 and 37 °C. Lane 1, control without both additive DNAs and Ce(ɪᴠ)/EDTA; lane 2, control in the absence of the additives [only with Ce(ɪᴠ)/EDTA]; lane 3, DNA$^{(L)}$-H/H-DNA$^{(R)}$ with Ce(ɪᴠ)/EDTA; lane 4, DNA$^{(L)}$-L$_{12}$-P/H-DNA$^{(R)}$ with Ce(ɪᴠ)/EDTA; lane 5, DNA$^{(L)}$-H/P-L$_{12}$-DNA$^{(R)}$ with Ce(ɪᴠ)/EDTA; lane 6, DNA$^{(L)}$-L$_{12}$-P/P-L$_{12}$-DNA$^{(R)}$ with Ce(ɪᴠ)/EDTA; M, the markers (authentic oligonucleotides having 3'-OH termini). Reaction conditions: [DNA$^{(S5)}$] = 1.0 μM, [each of the additive DNAs] = 2.0 μM, [NaCl] = 100 mM, and [Ce(ɪᴠ)/EDTA] = 1.0 mM at pH 7.0 (7.5 mM Hepes buffer) and 37 °C. (b) Magnified versions of the main part in lanes 4–6.

DNA, and the monophosphate group(s) is placed at the edge of these gaps. Figure 7.4 shows the typical results of polyacrylamide gel electrophoresis (PAGE) for the site-selective scission. In lane 4, DNA$^{(L)}$-L$_{12}$-P and H-DNA$^{(R)}$ are combined, and a 5-base gap is formed from T21 to G25 in the substrate DNA$^{(S5)}$. Thus, the monophosphate group is placed at the 5′-side edge of this 5-base gap. Upon treating this system with Ce(IV)/ EDTA complex at pH 7.0 and 37 °C, the gap-site is selectively hydrolyzed (lane 4). Scission occurs exactly in the gap-site and ranges from T21 to G25. Site-selective DNA scission is also successful when one monophosphate group is placed at the 3′-side edge of the gap (DNA$^{(L)}$-H/P-L$_{12}$-DNA$^{(R)}$ combination: lane 5). The DNA$^{(L)}$-L$_{12}$-P/P-L$_{12}$-DNA$^{(R)}$ combination, which provides two monophosphate groups to the gap, is still more effective (lane 6). Without the terminal monophosphate in the oligonucleotide additives, however, DNA scission by Ce(IV)/EDTA is far (>10-fold) slower (lane 3).

7.5.1.2 Effects of Gap-length on Site-selective DNA Scission

Figure 7.5(a) shows the dependencies of site-selectivity and scission efficiency on gap length (number of unpaired nucleotides). When a 3-base gap is formed from T21 to A 23 in DNA$^{(S3)}$ by use of the oligonucleotide additives bearing monophosphate, this gap-site is selectively hydrolyzed by Ce(IV)/EDTA (lane 3). Furthermore, site-selective hydrolysis of a 2-base gap (lane 2) or a 1-base gap (lane 1) is also successful. The total

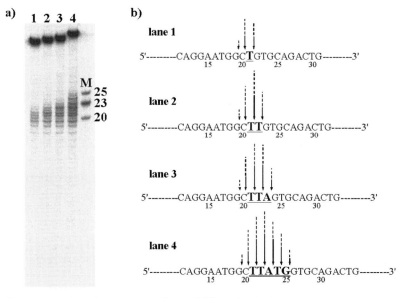

Figure 7.5 (a) Site-selective scission of gaps of different lengths by Ce(IV)/ EDTA at pH 7.0 and 37 °C. Lanes: 1, one-base gap; 2, two-base gap; 3, three-base gap; 4, five-base gap; M, the markers (authentic oligonucleotides having 3′-OH termini). These gaps were formed in DNA$^{(S1)}$-DNA$^{(S5)}$ (^{32}P-labelled at the 5′-end) using the DNA$^{(L)}$-L$_{12}$-P/P-L$_{12}$-DNA$^{(R)}$ combination. (b) Quantitative analysis of scission efficiencies (solid part is for the formation of 3′-OH terminus and the broken one for the formation of 3′-phosphate terminus).

conversion in the gap-region increases with increasing gap-length, since the number of scissile linkages increases.

In Figure 7.5(b), the relative scission-efficiency at each phosphodiester linkage is presented by the length of arrow (the solid parts and the broken parts correspond to the formation of 3'-OH and 3'-phosphate termini, respectively, vide infra). In all cases, the scissions are restricted strictly to the gap-site so that the site-selectivity increases monotonically with decreasing gap-length. With a 1-base gap, scission primarily takes place only at both sides of the unpaired nucleotide. Scission at the 2-base gap (as well as at the 3- and 5-gaps) is vigorous in the middle of the gap. Significantly, the efficiency for the scission of each linkage in the gap-site is not much dependent on either the gap-length or the sequence of the gap-site. Thus, the position for site-selective scission can be freely chosen. This is one of the most significant advantages of the present site-selective DNA scission.

7.5.1.3 Termini Structures of Scission Fragments

The scission fragments in Figures 7.4 and 7.5 (^{32}P-labelled at the 5'-end) are classified into two categories. One group co-migrates with authentic samples of oligonucleotides that have 3'-OH termini (in lane M). Conversely, another group is found between the bands of authentic samples. Apparently, the first group has 3'-OH termini and the second group has 3'-phosphate termini. In most cases, the 3'-OH termini [shown by the solid parts of the arrows in Figure 7.5(b)] are preferentially formed to the 3'-phosphate termini (broken parts). Similarly, 5'-termini of the fragments mostly have phosphate rather than OH termini. Thus, the present DNA hydrolysis mostly proceeds via P–O(3') scission of phosphodiester linkage, providing the fragments having the same terminus structures as formed by most of the naturally occurring nucleases. These fragments are easily ligated with other DNAs by enzymatic reactions (vide infra).

7.5.1.4 Essential Role of Gap-structure for the Site-selective DNA Scission

The gap structures are crucially important here, since the target site must be differentiated from other sites in terms of intrinsic reactivity. Thus, site-selective scission is unsuccessful when only one oligonucleotide additive bearing a monophosphate is used (without the second oligonucleotide additive). Here, no gap-structure is formed in substrate DNA, and the single-stranded portion of the DNA substrate is hydrolyzed by Ce(IV)/EDTA without any remarkable selectivity. Consistently, DNA scission at the target site is never promoted when oligonucleotides bearing only the linker groups (and no monophosphate groups) are used. The monophosphate groups are essential to recruit the Ce(IV) to the target site for site-selective scission. When a small gap is hydrolyzed by the present system, the scission efficiency depends significantly on the length of the linker between DNA and a phosphate group. For example, scission at a 1-base gap by the $DNA^{(L)}$-L_{12}-P/P-L_{12}-$DNA^{(R)}$ combination is about $3\times$ as fast as that by the $DNA^{(L)}$-L_0-P/P-L_0-$DNA^{(R)}$ combination. However, the efficiency of scission at a 3-base gap and a 5-base gap is less dependent on linker length.

7.5.2
Enzymatic Ligation of the Fragments Obtained by Site-selective Scission

The present site-selective scission proceeds by an hydrolytic pathway, as is the case in DNA scission by nucleases. Thus, the scission fragments can be recombined with various oligonucleotides by using DNA ligase. Figure 7.6 depicts a typical example. In the presence of the DNA$^{(L)}$-L$_0$-P/P-L$_0$-DNA$^{(R)}$ combination, DNA$^{(S5)}$ is first hydrolyzed by Ce(IV)/EDTA (lane 2), and the scission fragments are purified by PAGE. To this reaction mixture, an oligonucleotide having a monophosphate at the 5'-end (DNA$^{(ligated)}$: arbitrary sequence and length) and DNA$^{(template22)}$ template are added, and the mixture

a)

DNA$^{(S5)}$: 5'-FAM-CAATTAGAATCAGGAATGGC**TTATG**GTGCAGACTGTCGACCTAAG-3'

DNA$^{(template22)}$: 3'-CCTTACCGAA——CAGCACATAG-5'

DNA$^{(template24)}$: 3'-TTACCGAATA–CAGCACATAG-5'

DNA$^{(ligated)}$: 5'-pGTCGTGTATCCTAGGCTATGTGTAGTCTGCATGTACTGC-3'

b)

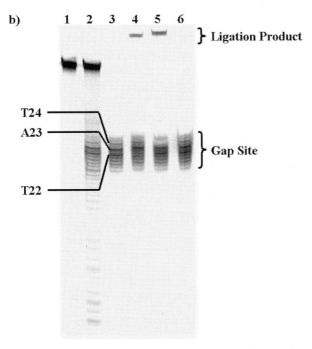

Figure 7.6 Ligation of the scission fragments by T4 DNA ligase in the presence of various templates. (a) Sequences of the oligonucleotides used. (b) Lane 1, DNA$^{(S5)}$ without treatment (labeled by fluorescein at the 5'-end); lane 2, product of site-selective scission of DNA$^{(S5)}$ by Ce(IV)/EDTA complex in the presence of DNA$^{(L)}$-L$_0$-P/P-L$_0$-DNA$^{(R)}$ combination; lane 3, product in lane 2 purified by PAGE (fragments ranging from C1 to T20-G26 were collected); lane 4, ligation product in the presence of DNA$^{(template22)}$; lane 5, ligation product in the presence of DNA$^{(template24)}$; lane 6, product obtained by ligation in the absence of templates.

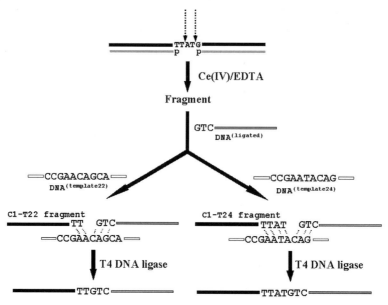

Figure 7.7 Schematic representation of enzymatic ligation of the scission fragment and foreign DNA using various templates.

is treated with T4 DNA ligase (lane 4). The 3′-side portion of DNA$^{(template22)}$ (10 nucleotides) is complementary with G13-T22 of DNA$^{(S5)}$, and the remainder of this template is complementary with the 5′-side of DNA$^{(ligated)}$ [Figure 7.6(a)]. Upon treatment with T4 DNA ligase, a new band appears at the top of the gel. This is the ligation product of the C1-T22 fragment, formed by cutting DNA$^{(S5)}$ between T22 and A23, and the DNA$^{(ligated)}$ (shown schematically on the left-hand side of Figure 7.7). The present DNA scission occurs at several linkages in DNA$^{(S5)}$, so that there exist several fragments in the reaction mixture. However, only one fragment that completely fits the template oligonucleotide is picked up and connected with DNA$^{(ligated)}$. Consistently, the band T22 for the C1-T22 fragment bearing 3′-OH terminus becomes weaker in the enzymatic reaction, while the intensities of other bands hardly change (compare lanes 3 and 4).

These arguments are further supported by the results in lane 5, where the DNA$^{(template24)}$ is used in place of DNA$^{(template22)}$. This template is complementary with A15-T24 of DNA$^{(S5)}$ and with the 5′-side of DNA$^{(ligated)}$. A new band of small mobility also appears after the treatment by DNA ligase, corresponding to the ligation product between the C1-T24 fragment and DNA$^{(ligated)}$ (see right-hand side of Figure 7.7). This product is two bases longer than the one formed in lane 4 and, thus, has a slightly smaller mobility. The C1-T24 fragment band almost disappears upon enzymatic treatment (cf. lanes 3 and 5). In the absence of templates, no ligation products are formed (lane 6). By designing the template oligonucleotide appropriately, the desired scission fragment is selectively incorporated to recombinant DNA. Thus, the present system is promising for future gene manipulation of single-stranded DNA (Figure 7.8).

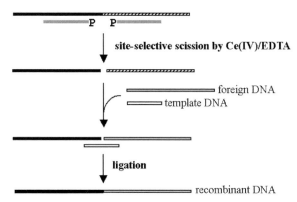

Figure 7.8 Proposed method for gene manipulation of single-stranded DNA.

7.6
Site-selective Scission of Double-stranded DNA by Combining Ce(ɪᴠ)/EDTA Complex with Pseudo-complementary PNA

7.6.1
Design of Artificial Restriction Enzymes for Double-stranded DNA Scission

Despite many attempts, there have been few reports on artificial restriction enzymes for site-selective hydrolysis of double-stranded DNA [15]. Since most of DNA in nature is double-stranded, these tools are quite important for practical applications. The following tactics, developed only recently, also take advantage of the remarkable substrate-specificity of the Ce(ɪᴠ)/EDTA complex (preferential hydrolysis of single-stranded DNA over double-stranded DNA). Thus, gap-like structures are formed in both strands of double-stranded DNA by using the invasion of pseudo-complementary PNA (pcPNA), and these gap-like sites are used as "hot spots" for selective hydrolysis by the complex. In pcPNA as DNA analog, phosphodiester linkages are replaced with amide linkages. Because of the absence of negative charges, pcPNA forms a more stable duplex with its complementary DNA than does the corresponding DNA (there is no electrostatic repulsion between negative charges) [16]. Furthermore, both adenine and thymine groups are chemically modified in pcPNA so that the formation of A-T base pairs in the duplex between two "complementary" pcPNAs is suppressed. Thus, the pcPNA/pcPNA duplex is only inefficiently formed, whereas formation of the pcPNA/DNA duplex is less affected by the chemical modification. Accordingly, pcPNA easily invades its complementary part in double-stranded DNA without much interference from the formation of pcPNA/pcPNA duplexes [17].

The two pcPNA additives used are designed so that a single-stranded portion is formed at predetermined sites in both strands of double-stranded DNA substrate when they invade the substrate; the structure is shown in the bottom of Figure 7.9(a). These gap-like sites are the "hot spots" that are selectively hydrolyzed by Ce(ɪᴠ)/EDTA, resulting in the site-selective scission of both strands at the target

a) Linearized PBR322 (4361 base pairs) and PNA additives

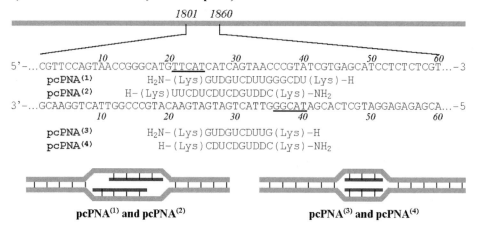

b) DNA fragments used for enzymatic ligation

Figure 7.9 (a) DNA substrate and pcPNA additives used for site-selective scission by Ce(IV)/EDTA. U and D in pcPNA sequences bear 2-thiouracil and 2,6-diaminopurine residues in place of conventional bases. In the invasion-complex composed of linearized PBR322, pcPNA(1), and pcPNA(2), the underlined nucleotides remain unpaired. (b) DNA fragments used for enzymatic ligation. The sticky end of foreign DNA (DNA(1)/DNA(2) duplex) is complementary with that of the scission fragment (fragment(L)).

site [18]. Furthermore, fragments obtained by the scission are successfully ligated with foreign DNA to provide recombinant DNA (Section 7.6.3). Of course, the sequences and lengths of pcPNA additives are freely chosen. Thus, the present method is, in principle, applicable to site-selective scission of larger DNA substrates, and facilitates the manipulation of these DNAs for future biotechnology.

7.6.2
Site-selective Hydrolysis of Double-stranded DNA

Both linear and supercoiled double-stranded DNA can be used as substrate for site-selective scission. Typical methods for the scission of linear double-stranded DNA are given in Sections 7.6.2.1 and 7.6.2.2. The substrate is obtained by treating supercoiled PBR322 plasmid DNA by a restriction enzyme *Eco*RI. Here, the plasmid DNA is cut at one-site, providing a linear DNA (this DNA is known as form III). Section 7.6.2.3 describes the method for site-selective hydrolysis of supercoiled DNA. Figure 7.9(a) shows the sequences of pcPNA additives (as well as those of DNA substrate near the scission sites).

7.6.2.1 Invasion of Two pcPNAs into Double-stranded DNA

The invasion complex was formed simply by incubating the linearized DNA with pcPNA$^{(1)}$ and pcPNA$^{(2)}$, as clearly evidenced by a gel shift assay. These two pcPNA additives are complementary with C1826-A1840 in the upper strand of the linearized DNA and A1821-G1835 in its lower strand, respectively [Figure 7.9(a)]. Accordingly, when these pcPNA additives invade the linearized DNA, a gap-like structure is formed in each strand [T1821-T1825 of the upper strand and G1836-T1840 in the lower strand: underlined bases in Figure 7.9(a)]. These gap-like (single-stranded) sites are the scission targets.

a) site-selective scission of linearized PBR322

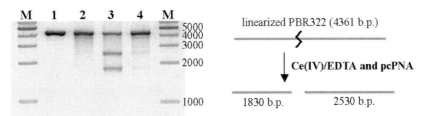

b) direct site-selective scission of supercoiled PBR322

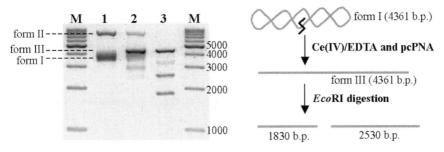

Figure 7.10 Agarose gel electrophoresis patterns for Ce(IV)/EDTA-induced site-selective hydrolysis of double-stranded DNA by Ce(IV)/EDTA and pcPNA additives. Bands were detected by staining with GelStar. (a) Site-selective hydrolysis of linearized PBR322 using pcPNAs. Lane 1, control; lane 2, Ce(IV)/ EDTA only; lane 3, pcPNA$^{(1)}$/pcPNA$^{(2)}$ + Ce(IV)/EDTA; lane 4, pcPNA$^{(3)}$/ pcPNA$^{(4)}$ + Ce(IV)/EDTA; M, 1000 base pair ladder. [linearized PBR322] = 4 nm, [each of pcPNAs] = 100 nm, [Ce(IV)/EDTA] = 20 μM, [NaCl] = 100 mm, at pH 7.0 (5.0 mm Hepes buffer) and 37 °C. (b) Direct site-selective hydrolysis of supercoiled PBR322. Lane 1, control; lane 2, pcPNA$^{(1)}$/pcPNA$^{(2)}$ + Ce(IV)/EDTA; lane 3, *Eco*RI digests of the products in lane 2.

7.6.2.2 Site-selective Scission

When the invasion complex composed of linearized PBR322, pcPNA[1], and pcPNA[2] was treated with Ce(IV)/EDTA complex at pH 7.0 and 37 °C, two fragments were formed, as observed in agarose gel electrophoresis [lane 3 in Figure 7.10(a)]. One fragment is between 2000 and 3000-mer, and the other is a little smaller than 2000-mer. When scission occurred at the target gap-like sites, products of ca. 1830 and 2530-mer should be formed from the left- and right-hand sides of the substrate DNA, respectively. Apparently, the Ce(IV)/EDTA complex hydrolyzed selectively both strands exactly as designed.

For the present site-selective scission of double-stranded DNA, both pcPNA additives must have flanking portions so that gap-like structures are formed in the double-stranded DNA. Thus, no site-selective scission was observed when pcPNA[3] and pcPNA[4] were combined as additives [Figure 7.10(a), lane 4]. These two pcPNAs are completely complementary (their duplex is not much formed due to mutual steric repulsion under the conditions employed), and thus no gap-like structures are produced in the invasion complex [see structure at the bottom of Figure 7.9(a)]. No scission occurred in the absence of pcPNA additives, as expected (lane 2).

7.6.2.3 Direct Site-selective Scission of Supercoiled Plasmid DNA

In virtually the same way, supercoiled DNA can be hydrolyzed site-selectively by Ce(IV)/EDTA [Figure 7.10(b)]. The invasion complex was obtained by incubating PBR322 plasmid DNA with pcPNA[1] and pcPNA[2], and was then treated with the Ce(IV) complex. In this treatment, the supercoiled DNA was converted into linearized DNA (form III). When the product was digested by the restriction enzyme *Eco*RI (to analyze the scission site), two fragments of expected sizes [the same ones as obtained in lane 3 of Figure 7.10(a)] were formed [lane 3 in Figure 7.10(b)]. Apparently, the scission of the supercoiled DNA took place at the same site as that of the linearlized DNA. The present site-selective scission is applicable to various DNAs.

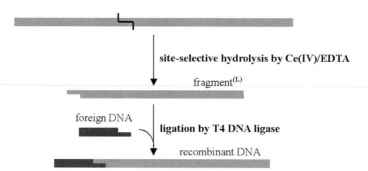

Figure 7.11 Schematic representation for manipulation of double-stranded DNA using the present artificial restriction enzyme.

Figure 7.12 Agarose gel electrophoresis of the PCR product from the recombinant DNA. Lane 1, PCR product from the recombinant DNA; lane 2, PCR product from the mixture without the enzymatic ligation; M, 200 base pair ladder.

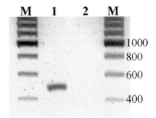

7.6.3
Enzymatic Ligation of the Scission Fragment and Foreign DNA

The scission products were treated with foreign double-stranded DNA in the presence of T4 DNA ligase (Figure 7.11). For example, the fragment obtained in Figure 7.10(a) (the one between 2000 and 3000-mer) was combined with DNA[(1)]/DNA[(2)] duplex. The sticky end of this foreign DNA is complementary with the end of the fragment[(L)] that is formed when the upper strand of the linearized plasmid DNA is hydrolyzed at the 5′-side of A1824 and the lower strand is cleaved at the 5′-side of G1837 [see Figure 7.9(b)].

Successful, selective ligation of these two double-stranded DNAs has been substantiated by the following experiments. The ligation product was first amplified by PCR. Of the two primers used, primer[(1)] is complementary with DNA[(2)], whereas primer[(2)] is complementary with G2271-C2292 of the upper strand of the fragment[(L)]. As shown in lane 1 of Figure 7.12, a PCR product of about 500-mer size was formed. When these two double-stranded DNAs were successfully ligated and the PCR proceeded with the ligation product as the template, 492-mer PCR product should be formed. Conversely, this PCR product should not be formed if the ligation was unsuccessful. Consistently, no PCR products were formed when a mixture of the longer scission products (including the fragment[(L)]) and DNA[(1)]/DNA[(2)] duplex was directly (without treatment with T4 ligase) used for the PCR (lane 2, Figure 7.12). Furthermore, the PCR product obtained above by using primer[(1)] and primer[(2)] was characterized completely by sequencing experiments (Figure 7.13).

In the scission of double-stranded DNA by Ce(IV)/EDTA in the presence of pcPNA[(1)]/pcPNA[(2)], several fragments should be formed. In the enzymatic ligation, however, the scission fragment (fragment[(L)] here), whose sticky end completely fits

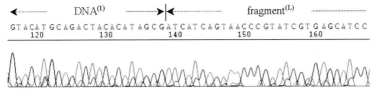

Figure 7.13 Result of sequencing analysis of the recombinant DNA. The sequence involving DNA[(1)] and the 5′-side portion of upper strand of the fragment[(L)] is shown.

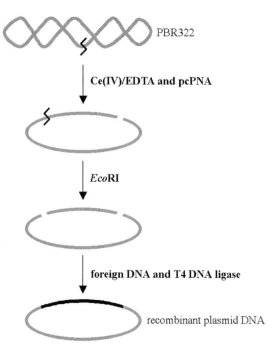

Figure 7.14 Preparation of recombinant plasmid DNA using Ce(IV)/EDTA and pcPNA additives.

that of foreign DNA (DNA$^{(1)}$/DNA$^{(2)}$ duplex), is picked up and selectively incorporated into the recombinant DNA. Furthermore, recombinant plasmid DNA has been successfully prepared using the present artificial restriction enzymes (Figure 7.14). Since pcPNA can invade DNAs of various sequences, these artificial restriction enzymes should be applicable to the manipulation of still longer DNAs.

7.7
Conclusion

The preparation of artificial restriction enzymes has been one of the most attractive themes for chemists and biochemists, since they are essential for manipulating huge DNA in future biotechnology. As described in this chapter, useful tools for site-selective scission of either single-stranded or double-stranded DNA are now in hand. The key factor for the molecular design of these artificial enzymes is preferential hydrolysis of single-stranded portions by the Ce(IV)/EDTA complex over double-stranded portions. Accordingly, substrate DNA is site-selectively hydrolyzed when a gap-structure (or its relevance) is formed at a predetermined position by using appropriate DNA or PNA additives. The sequences and lengths of these additives can be freely chosen, and thus these systems are applicable to DNA substrates of any length and sequence.

Furthermore, as the present site-selective scission proceeds completely via a hydrolytic pathway, the scission fragments can be recombined with foreign DNA to provide desired recombinant DNA. The fundamental processes of DNA manipulation (site-selective scission of DNA, followed by enzymatic ligation) are successfully accomplished. The methods presented here are satisfactorily compatible with current molecular biology and biotechnology, and thus should be highly promising for future applications. Such possibilities, as well as attempts to further improve scission-efficiency and -selectivity, are currently being pursued in our laboratory.

Acknowledgment

This work was partially supported by PROBRAIN. Support by a Grant-in-Aid for Scientific Research from the Ministry of Education, Science, Sports, Culture and Technology, Japan is also acknowledged.

References

1 Reviews on relevant topics: (a) M. Komiyama, J. Sumaoka, *Curr. Opin. Chem. Biol.* **1998**, *2*, 751–757. (b) E. L. Hegg, J. N. Burstyn, *Coord. Chem. Rev.* **1998**, *173*, 133–165. (c) N. H. Williams, B. Takasaki, M. Wall, J. Chin, *Acc. Chem. Res.* **1999**, *32*, 485–493. (d) R. Ott, R. Krämer, *Appl. Microbiol. Biotechnol.* **1999**, *52*, 761–767. (e) A. Sreedhara, J. A. Cowan, *J. Biol. Inorg. Chem.* **2001**, *6*, 337–347. (f) S. J. Franklin, *Curr. Opin. Chem. Biol.* **2001**, *5*, 201–208. (g) C. Liu, M. Wang, T. Zhang, H. Sun, *Coord. Chem. Rev.* **2004**, *248*, 147–168.

2 J. Chin, M. Banaszczyk, V. Jubian, X. Zou, *J. Am. Chem. Soc.* **1989**, *111*, 186–190.

3 (a) Y. Matsumoto, M. Komiyama, *Chem. Express* **1992**, *7*, 785–788. (b) Y. Matsumoto, M. Komiyama, *Nucl. Acids, Symp. Ser.* **1992**, *27*, 33–34.

4 M. Komiyama, K. Matsumura, Y. Matsumoto, *J. Chem. Soc., Chem. Commun.* **1992**, 640–641.

5 M. Komiyama, N. Takeda, Y. Takahashi, H. Uchida, T. Shiiba, T. Kodama, M. Yashiro, *J. Chem. Soc., Perkin Trans. 2* **1995**, 269–274.

6 J. Sumaoka, Y. Azuma, M. Komiyama, *Chem. Eur. J.* **1998**, *4*, 205–209.

7 T. Ihara, H. Shimura, K. Ohmori, H. Tsuji, J. Takeuchi, M. Takagi, *Chem. Lett.* **1996**, *25*, 687–688.

8 R. Ott, R. Krämer, *Angew. Chem., Int. Ed.* **1998**, *37*, 1957–1960.

9 (a) M. Komiyama, T. Shiiba, Y. Takahashi, N. Takeda, K. Matsumura, T. Kodama, *Supramol. Chem.* **1994**, *4*, 31–34. (b) M. Komiyama, *J. Biochem.* **1995**, *118*, 665–670.

10 F. H. Zelder, A. A. Mokhir, R. Krämer, *Inorg. Chem.* **2003**, *42*, 8618–8620.

11 Non-covalent strategy has been used also for site-selective scission of RNA: (a) D. Hüsken, G. Goodall, M. J. J. Blommers, W. Jahnke, J. Hall, R. Häner, H. E. Moser, *Biochemistry* **1996**, *35*, 16591–16600. (b) A. Kuzuya, R. Mizoguchi, F. Morisawa, K. Machida, M. Komiyama, *J. Am. Chem. Soc.* **2002**, *124*, 6887–6894.

12 T. Igawa, J. Sumaoka, M. Komiyama, *Chem. Lett.* **2000**, 356–357.

13 (a) Y. Kitamura, M. Komiyama, *Nucl. Acids Res.* **2002**, *30*, e102. (b) Y. Kitamura, J. Sumaoka, M. Komiyama, *Tetrahedron* **2003**, *59*, 10403–10408.

14 (a) W. Chen, T. Igawa, J. Sumaoka, M. Komiyama, *Chem. Lett.* **2004**, *33*, 300–301. (b) W. Chen, Y. Kitamura, J.-M. Zhou, J. Sumaoka, M. Komiyama, *J. Am. Chem. Soc.* **2004**, *126*, 10285–10291.

15 Conjugation of metal-binding protein with DNA-binding protein is one solution: R. T. Kovacic, J. T. Welch, S. J. Franklin, *J. Am. Chem. Soc.* **2003**, *125*, 6656–6662.

16 M. Egholm, O. Buchardt, L. Christensen, C. Behrens, S. M. Freier, D. A. Driver, R. H. Berg, S. K. Kim, B. Norden, P. E. Nielsen, *Nature* **1993**, *365*, 566–568.

17 J. Lohse, O. Dahl, P. E. Nielsen, *Proc. Natl. Acad. Sci. U.S.A.* **1999**, *96*, 11804–11808.

18 (a) Y. Yamamoto, M. Komiyama, *Chem. Lett.* **2004**, *33*, 920–921. (b) Y. Yamamoto, A. Uehara, T. Tomita, M. Komiyama, *Nucleic Acids Res.* **2004**, *32*, e153.

Index

Artificial Enzymes. Edited by R. Breslow
Copyright © 2005 WILEY-VCH Verlag GmbH & Co. KGaA, Weinheim
ISBN: 3-527-31165-3